The new book Mysticism & Physics

By

Angelo Aulisa

Copyright 2018 by Angelo Aulisa

First edition March 2018

ISBN

ALL RIGHT RESERVED

No part of this publication may be reproduced in any form or by any means electronic or mechanical

Including photocopying recording or any information browsing storage or retrieval system without

Permission in writing form from the author or publisher

Published by Angelo Aulisa

Distributed to the trade by ....

The new book Mysticism & Physics Table of contents Angelo Aulisa

Chapter 1 The invisible tread that entangled particles here

Chapter 2 and you sing la cicirignola

Chapter 3 Anthropomorphism

Chapter 4 Anthropomorphism part two

Chapter 5 Consciousness & BBC

Chapter 6 Consciousness & BBC

Chapter 7 History of the universe comment

Chapter 8 History of the universe comment part two

Chapter 9 Unconscious & conscious

Chapter 10 Unconscious & conscious part two

Chapter 11 Unconscious & conscious part 3

Chapter 12 Congratulation to Nobel price

Chapter 13 Congratulation to Nobel price part two

Chapter 14 Woman & enlightenment

Chapter 15 Woman & enlightenment part two

Chapter 16 Woman & enlightenment part 3

Chapter 17 Woman & enlightenment part 4

Chapter 18 Melt down & new dawn of consciousness

Chapter 19 Melt down & new dawn of consciousness

Chapter 20 Doing things together

Chapter 21 Doing things together part two

Chapter 22 Seven astral body of an organic unity

Chapter 23 Seven astral body of an organic unity part two

Chapter 24 Seven astral body of an organic unity part 3

Chapter 25 Seven astral body of an organic unity part 4

Chapter 26 Seven astral body of an organic unity part 5

Chapter 27 Sabbath is made for people & not people for Sabbath

Chapter 28 Sabbath is made for people & not people for Sabbath part two

Chapter 29 Sabbath is made for people & not people for Sabbath part 3

Chapter 30 Meditation & path

Chapter 31 Meditation & path part two

Chapter 32 Meditation & path part 3

Chapter 33 Meditation & path part 4

Chapter 34 Generation gap

Chapter 35 Generation gap part two

Chapter 36 Generation gap part 4

Chapter 37 Generation gap part 3

Chapter 38 Generation gap part 5

Chapter 39 Generation gap part 6

Chapter 40 Generation gap part 7

Chapter 41 toward civilization

Chapter 42 toward civilization part two

Chapter 43 toward civilization part 3

Chapter 44 toward civilization part 4

Chapter 45 toward civilization part 5

Chapter 46 toward civilization part 6

Chapter 47 toward civilization part 7

Chapter 48 toward civilization part 8

Chapter 49 toward civilization part 9

Chapter 50 toward civilization part 10

Chapter 51 toward civilization part 11

Chapter 52 toward civilization part 12

Chapter 53 toward civilization part 13

Chapter 54 toward civilization part 14

Chapter 55 toward civilization part 15

Chapter 56 toward civilization part 16

Chapter 57 toward civilization part 17

Chapter 58 toward civilization part 18

The new book Mysticism & Physics Introduction Angelo Aulisa

Hi friends , welcome to my new book on mysticism & physics one of the most beautiful book ever write in the history of humanity , which out date all the books that have been write down the age of the history of humanity , it out date those books because this book come to light in a contemporary of 2018 , and the author Angelo Aulisa is a man of our age and contemporary, and use all the tools of mysticism physics sciences , psychology , psychoanalysis , literature , and all the evolution that humanity as gone through , such as convergence of evolution of DNA that as had thousands of mutation convergence of evolution in better shape of living and intelligence in better quality of living ,our DNA aspire always and mutate always in accordance with the new situation it change shape , quality in order to be contemporary to the new situation and survive , but not only our DNA aspire to a final refinement of itself into an ultimate convergence of evolution mutation which are immortality and resurrection , same those our inner consciousness it aspire always in a propulsion evolution toward a final refinement of itself, into light empty consciousness , and finally annihilate into non being body incorporeal were time , space forms duality of mind complete annihilate, and consciousness annihilate or better it twisted turn into relation less unfocused awareness, that is just an I am ness a great relaxation into the law of eternity , begin less endless never born or die , that the meaning of eternity , awareness is the ultimate essence of the law of eternity , were immortality and resurrection happen unfold , eternity is the ultimate canvas of the universal body bigger huger vaster than the universe itself , and beyond above transcendental to the universe itself , actually is the ultimate canvas were the universe is display paint , and is an infinite freedom without boundary , infinity like the ultimate equation of quantum mathematics , no size is just an infinite open relativity not absolute at all because it begin nowhere and it end nowhere sacred divine holy , is just infinite silence intrinsic with subtle ecstasy , an infinite bliss , eternity is the core and source of the mystery of the universe and of life and death and of duality , into eternity you are at home and is were immortality and resurrection eternal unfold , the path to it meditation silence love , or any path that make you mindless centered into your inner being , singing painting dancing , sculpting , running , etc.. . Into the new book of mysticism & physics the divine melody sacred divine is spelling with such accuracy that you will be shock , surprise , in wonder , and it will bring into the reader a new update consciousness awareness , a new transformation of is life that you have not even imagine before or dream about before . The new book of mysticism & physics basically it regard mysticism , which means the inner science of the inner reality of men , such as inner being , inner consciousness , non being body incorporeal , inner awareness into the law of eternity , the path to it per excellence is meditation done for real , but also silence love or any path that take you into a mindless dimension less dimension or into the present moment here now flowing within your inner being and consciousness , is a search research of your intrinsic original nature , love by the way , that the meaning of mysticism an inner search of the inner reality of men call it inner science or mysticism right . You will enjoy this inner journey as never before in your life and it will transform you complete chapter by chapter, you will get new insight into life and death , and most of all at the end of the book if you have the patient to read meditative all of the book you will be enlightened conscious aware , awake from unconscious sleep forever and ever , a new fresh begin of your life this time authentic real scientific factual for your day to day life , with an update consciousness awareness to our contemporary 2018 , and you will drop forever old traditions expire and out of date old religions expire and out of date a night mare of the past ancient humanity lies and fancy fairy tale for retarded children , that were hindering your evolution in intelligence and consciousness , they were simple castrate you spiritually , I don't like the term spiritually, I use the term conscious is more update accurate they were castrate you for your conscious evolution in intelligence and universal consciousness , weak up get smart and let go into the transformation evolution high light into this fantastic book , you will be reborn as nature as made you originally .And this book regard Physics , from ancient Greek means the knowledge of nature , from Greek Physic means nature , is the natural science that studies matter and is motion behavior through space and time , and studies the related entities of energy and forces , physics is one the most fundamental disciplines , and is main goal is to understand how the universe behaves . Physics one of the oldest academic disciplines , and through is inclusion of Astrology perhaps the oldest over the last two millennia , physics , chemistry biology and certain branch of mathematics were part of natural philosophy , but during the scientific revolution in the 17 century these natural sciences emerge as unique research endeavors in their own right , physics intersects with many interdisciplinary areas of research as biophysics and quantum chemistry , and the boundary of physics are not rigidly define . New ideas in physics often explain the fundamental mechanisms studies by other sciences   , and suggest new avenues of research in academic disciplines such as mathematics and philosophy. Theory Cal   breakthrough in physics also make significant contribution by enabling advances in new technologies , for example advances in the understanding of electromagnetism and nuclear physics led directly to the development of new product that dramatically transformed modern day society , such as television computers , domestic appliances , and nuclear weapons , advances in thermodynamics led to the development of industrialization and advances in mechanics inspire the development of calculus . The physics of modern day as become meta physics and merge converge with mysticism in many circumstance, and in this book you will find all the convergence of physics with mysticism, and what emerge is a synthesis unique and amazing never witness before from human kind, you are all welcome for this tremendous quantum leap mutation convergence of evolution that you will get through reading this book of mysticism & physics. I want added that mysticism is not science were you get a formula and all can benefit from it , no in mysticism they are no formula , everyone as to go for the experience alone and each experience will differ from one another , what Mahavira experience is unique to himself , what Jesus experience is unique to himself , what Lao-tzu experience is unique to himself , what you will experience is unique to yourself , everyone as to be the begin of new religion less religion, in short this is the theory of relativity, what was good and beneficial for Jesus maybe a big hindrance for you and what was good for all Prophet may be a big hindrance for their follower , everyone is unique as is intrinsic quality attitude , you see for your chief quality attitude and follow originally don't be a parrot emulator copy , it do not function work , be yourself originally and it work function , they are no formula your experience experimentation is determinate absolutely require , so let go, on the path of meditation love silence, and experience on your own and you are in for the greatest surprise of your life , enlightenment will be yours , remember this freedom means literally responsibility each of your action you are responsible , there is nobody above you , and you are the only responsible of your life , you born alone you live alone if it happen to share love wisdom experience welcome it don't be miserable let go and go for it be abundant life is abundant , and you die alone finally , freedom is responsibility of your own life , no govern no institution no community is responsible for your life , take the responsibility on your own and build a life of quality of freedom of playfulness sacred holy divine ……Angelo Aulisa

The new book Mysticism & Physics
By
Angelo Aulisa

The new book Mysticism & Physics Chapter 1 Angelo Aulisa

The invisible tread that entangled particles here

Hi friends , what is the invisible tread that entangled two particles when they are far distant into the universe , that is speed then light even , .To me is very simple , Consciousness is everywhere and nowhere in particular expand into the universe , is the inner tread that run through all living being and forms of the universe , which is a fundamental law of the universe a pulsation of love intelligence , a quality a creativity , simultaneous to the whole universe like an invisible canvas but flowing alive is the within intelligence of the universe , the universe is very intelligent irrational but absolutely intelligent the essence of intelligence the data of intelligence are running through the universe intrinsic to consciousness , I repeat fundamental law of the universe , that modern physic call it Boson x the quality that hold together keep together the all universe and confer Mass to particles atoms and matter , without the Boson x the whole universe disintegrate instant forms and the universe itself disintegrate instant , in mysticism the Boson x is called consciousness since thousands of years , and is the inner tread intrinsic to the universe is a conscious alchemy not physical not gross not material , and is neutral to Gender color race age , anyone through silence meditation , love can rediscover consciousness and be one with it , consciousness is label less content less adjective less just a pulsation quality creativity intrinsic to the universe flowing through all living being and forms of the universe , vector of the intelligence of the essence , of the fragrance of the data of the all Buddha awakened one enlightened one ever exist in the past in the present and in future too , .in fact into an atoms the are protons neutrons and electrons, that represent well this 3 tense of time , wen u are annihilate into consciousness this 3 tense unfold simultaneous , the past is no more is dead the present is now elusive flowing the future is not yet just imagination but they are contain within the atoms all 3 , and into consciousness all 3 tense unfold simultaneous in the singularity of the atoms in the singularity of the slpet second of the moment , and so what is the hide reality that entangled two particles that are far distant between them into the universe, and they communicate simultaneous in synchronicity and they maintain all property , that it seems that this hide reality is speed even of light , such an easy quest , is CONSCIOUSNESS that is simultaneous to the whole universe intrinsic to all living being and forms of it of the universe , and how the entangled particles preserve all property the insight the data that create the entanglement , is because consciousness is intrinsic to the particles entangled , and is eloquent the fact that entanglement of particles is eternal , two particles that get entangled they remain entangled even after death , the body die the forms objective die but consciousness is eternal never die , so even after death two particles that have being really entangled they continue to be entangled , in a way we in essence we can communicate remain entangled with the passed away , with the present one , and with the future one two , entanglement is a phenomenon that pierce past life , going deep into this phenomenon of entanglement we can uncover past life easy , consciousness is a fundamental law of the universe a quality a pulsation a creativity but is also an evolutionary process of refinement ,, it refine itself self on and on through the many life forms that goes through, until a total refinement , into light empty consciousness then it reach is last life into this particular subject that as refine totally is consciousness and is called enlightenment , wen consciousness is totally refine into the subject it get enlightened , and when the subject leave the body it annihilate blissful in peace into non being into the law of eternity , no begin no end never born or die that is eternity the ultimate canvas of the universe the core and source of the mystery of the universe and of life and death and of all duality , and this final annihilation is called resurrection eternal resurrection , which is a conscious alchemy not physical not gross not material and for eternity to come it flow in essence in fragrance in data of intelligence intrinsic to the universal body and consciousness is the flowing vector of the essence of the fragrance of the intelligence data of the refine consciousness that as annihilate into eternity this is called the phenomenon of resurrection and immortality , and it will have no more reincarnation that will be the last body that as gone through and as refine itself totally into light empty consciousness , and is disentangled from any other being or particles or subject the mysterious phenomenon of entanglement ceases , But until reincarnation take place the phenomenon of entanglement it will take place , two being that get really entangled into previsions life they remain entangled forever , entanglement is a mysterious phenomenon that transcend life and death because is incorporeal a conscious phenomenon mysterious but all together conscious of consciousness that is the vector that connect link the particles far distant from each other into universe in a simultaneous amazing way , If you want entanglement is a process and phenomenon mysterious that confirm the evolution of the universe that the universe is an organism alive intelligent that evolve mutate that as convergence of evolution and mutation all the time that as quantum leap all the time, towards better intelligence better quality and better shape of it , at infinitum , just like the DNA of a human being , that as all the time convergence of evolution mutation quantum leap towards better intelligence better quality of living better shape of living , same is the universe the universe is an infinite organism the human being is an organic unity but all together a miniature universe , Friends the universe the world as not a fix creator that create the universe the world in six days and then he went in vacation and nobody as seen him anymore , if it was so life as no possibility of any evolution , because the hypothetic God that create such a world would have create a perfect world and perfection is dead static , no to me the world the universe is just creativity , pulsation of love intelligence, quality , intrinsic to the universe CONSCIOUSNESS , a pulsation that easy relate two particle entangled together even if they are wide distant between them , and beautiful it preserve the property the essence of the entanglement that as happen and beautiful it communicate into the present simultaneous between the two particles , is not an invisible influence unknown that as more speed of the light , I laugh smile at this , is simultaneous communication because consciousness is divine sacred everywhere and nowhere in particular expand into the universe , hence I think here and here at millions of light years is heard simultaneous , and receive the news , and more beautiful that if one of the particles is observe star at , it collapsed the within property vanish and simultaneous the other particles far distant millions of light years also collapsed instant , because the intrinsic nature of consciousness is freedom and as you interfere with the freedom of the single particles either it change behavior or it collapsed , this is divine in my own understand , and more consciousness exist until the horizon of event of the universe , consciousness come to be simultaneous with the universe and is flowing phenomenon , consciousness come to be in relation to the universe and consciousness need always and always a relation to be , when the universe after big bang come to be consciousness in relation to the universe come to be and unfold expand into the universe everywhere and nowhere in particular , is the Boson x in physicist terms it confer Mass to the particles that constitute the material part of the universe and keep hold the universe together , in mystical union into an enchanting organism sacred divine , but beyond above the universe on the overlapping with non being body which u can call also the law of eternity , into nothingness emptiness gate less gate of the universe consciousness annihilate cease to be because into non being body time space forms duality totally annihilate and consciousness as no more relation to be and it twisted into unfocused awareness ultimate essence of eternity or of non being body , awareness is just an I am ness expand everywhere and nowhere in particular into eternity and is unfocused and relation less , is just a great relaxation an I am ness, a bliss , sacred divine a subtle ecstasy within the infinite silence of eternity and is where the resurrection immortality take place unfold intrinsic to eternity, is where the immortality reside , into the law of eternity an open relativity an infinite freedom without boundary a bliss a peace a silence full of subtle ecstasy sacred and divine that surpass all understand , Anyhow I love that physic quantum mathematics , quantum physic is entangled with this immense interesting mystery of life and death , I also aim fascinate at how far physic is reaching everyday closer and close to the core and source of the mystery of the universe, that is eternity itself , the core and source of the mystery of the universe and of life and death and of all duality , is amazing , I welcome this search that mingled merge physic with mysticism the science of the inner mystery of human being ....great ...thank you and welcome ....by love all of you friends ...Angelo Aulisa

The new book Mysticism & Physics   Chapter 2 Angelo Aulisa

And then what...you sing.la. Cicirignola...

Friends , this is the greatest Zen koan , you know Italian they have one thousand and one injustice deception but into the living country they are amazing beautiful people , you sing la Cicirignola , means you sing the cicada , ha is divine , sacred , wen in deep meditation the cicada start to sing and pierce the button of your mind , like water flowing and shape out the rocks , yes then you sing the cicada , so friends consciousness the fundamental law of the universe , intrinsic to it is just a pulsation of love intelligence, a quality a creativity , that run through the all living being and forms of the universe , neutral to any interpretation of the little men , label less content less , adjective less just an empty light consciousness , divine sacred , a bliss , an intelligence , an inner wisdom of the universe eternal , is the inner tread that create the interaction of entanglement of two particles far distant from each other , because is everywhere and nowhere in particular expand and simultaneous into the whole universe , and consciousness to exist need always and always a relation , of subject or object , you experience in your life you become conscious always in relation to an object or subject , otherwise consciousness is dispose and turn into twisted into awareness , unfocused and relation less , because awareness is just an I am ness , a great relaxation into non being eternity , where forms , time space , duality totally annihilate, and awareness is the ultimate relaxation , many people when they have no subject or object to be conscious about they become restless , frustrate , in anguish and agony , because they have nowhere to put the attention , and they have never experience meditation , they know nothing of inner being and inner reality of men , they pretend to have always a subject or object to give attention or to be conscious about , but they are a lot of empty gape where there is just silence emptiness nothingness , those are the moment of awareness unfocused relation less, moment of great relaxation , in those moment consciousness twisted turn into awareness , during the day they are infinite moment of silence of empty gape that if you have a little experience of meditation those are the moment when you could have relax totally into the core and source of the mystery of the universe and of life and death and of all duality ,into eternity itself, and you could have a great peace silence regeneration recharge, rejuvenated, healing of all your organic unity organs , brain you give a rest to your brain that is constant working 24 hours , in the day you think constant in a mentasm , in the night you dream constant the brain is over used , give a rest , it need a rest , into the gape of silence of awareness a great relaxation happen into the core and source , of the mystery of the universe , of course you should have at priory an experience of meditation and know your inner being your inner consciousness and awareness , , is a great secret that will change totally your life , wen consciousness is on you are conscious of a subject or object and wen consciousness is dispose not having any relation anymore it turn twisted into awareness unfocused relation less and ultimate essence of non being and eternity itself, , and you relax no thought no emotion no unconscious stuff no mind , but just a great HA the HA experience that the psychologist talk about , a great relaxation into the source and core of life and death eternity itself, and you recharge regenerate rejuvenated heal all your organic unity , they are plenty of moment like this during the day , no need to fall into anguish frustration agony restless , relax , those gape of silence are sacred divine those gape of emptiness are a divine peace , meditation , during the day is almost a continues cycle on and off of consciousness of being conscious and of into awareness and of and on into consciousness and be conscious again and again , consciousness wen as no relation of subject or object it turn of spontaneous natural, and spontaneous natural awareness is on , I know the western mind is restless a mentasm stress out all the time but be a little awake , and then the benediction wen consciousness is off is a bliss a peace a pin drop silence full of subtle ecstasy is divine , you are at home into the core and source of the law of eternity , into an open relativity into infinite freedom into a blissful sacred silence full of ecstasy subtle ecstasy and then canti la cicirignola you sing the cicada Ha this Ha this thousand time HA this ...Any how I start this post to write first of all about the inner tread that connect link two particles into an entanglement even if they far distant from each other , friends is consciousness the particles intrinsic to them is consciousness , consciousness in relation to the particles is conscious into the particles and the far distant particles entangled with the other particles intrinsic to them is consciousness and is conscious , the two particles start interacting in relation to each other because that is the nature of consciousness in relation of the subject object of the particles become conscious , and the interaction is just a natural consequence of consciousness the two particles become relate through the tread of consciousness and simultaneous even if distant millions of light years the communicate beyond time and space , here I think and here is heard millions of light years distant , is the subject of the particles that create the possibility for consciousness to relate each other , if the particles collapse by observation or staring immediately consciousness cease to be relate to the particles because the organism die collapsed , or it change behavior , the particles , because freedom is the basic substratum of consciousness and if you interfere with freedom immediate you create the collapse or changing of behavior , even in life into the human being any interference with the freedom and spontaneity of the human being create immediate conditioning and repression and wounds into the unconscious , such a soft and delicate is the nature of consciousness and life , now the human being do not collapsed but it has a safety measure called unconscious were he throw all the interference with is freedom , but the unconscious as limited capability of stuffing it , at a certain point it become saturated and stem is create release , like a cooking pot , That is it one in life should experience meditation is basic important to have a life in balance harmony with the organism of the universe , we live into an organism called cosmos which literally means harmony , and we are not living into a mechanism a sum of part put together and the human being is an organic unity and not a machine or robot , so a little acquaintance with meditations basic important , for a life in harmony balance peace , forget your out of date expire religions they are all of them a bunch of neurosis a night mare of the ancient past humanity primitive and unconscious , and get update your inner being and consciousness and awareness to our contemporary age 2018 , learn meditation , the link that connect you to the universal consciousness awareness non being eternity itself , the core and source of the mystery of the universe and of life and death and of all duality , that will give a life of peace of harmony authentic and real to your true nature of your inner being and your inner consciousness and awareness , silence and love are also path to it or dance , I am a dancer , is also a path to it or laugh is also a path to it wen u laugh u cannot think and is a joy , a bliss an ecstasy .... Ok Angelo Aulisa ....

Published on June 22, 2017 04:10 • 61 views

The new book Mysticism & Physics   Chapter 3   Angelo Aulisa

Anthropomorphism

Hi friends , first hello to all community of goodreads.com , which within me I love for their contribution to the world of knowledge wisdom of course theory through books , but still is a great contribution to freedom of speech and to make the world a better place , to me freedom of speech means civilization and simultaneous communication worldwide means to be contemporary to our age 2018 where evolution in intelligence and science technology has reach is apex ,and is a pleasure to be contemporary to our age and time . I was thinking to stop to quit of writing post books , because of my failure all the time to have any record or success , but in the end it prevail the love for humanity , it do not matter if I have no great success what matter is if even a single individual understand what I mean the significance of my discourses I have fulfill my task and is worthy going on write , by the way official I have no success no payment is coming sometime Amazon send a little payment nothing just few coins, but is a sign at least one two person read what I write , I am happy wen I receive the notification of Amazon I laugh , not for the money I don't care at all but is a sign of one person as read what I say and pay for it . I think on the web nobody pay for reading the books they all download gratis I know, is miss fortune for the author, but fortunate in the end I write out of passion and love for humanity, so I will go on forever, by the way the blog post on good reads always 50, 70 people they read so you see is good to go on write, as far I enjoy doing it I will go on and that is the case right now. So the post of today is about anthropomorphism because is a very important subject this humanity this so call civilization is afflict by this disease since thousands of years , Anthropomorphism is when someone think of God as a person, basically when someone personalize God , but to be more accurate in the vocabulary anthropomorphism means the showing or treating of animals Gods and objects as if they are human in appearance, character behavior , That is it , is a serious disease , because this humanity is thousands of years that goes on showing treating Gods as person s , is a very childish interpretation to show God in appearance character behavior as a person that make human being infantile retarded in sort pathological neurotic , this disease make human being (spiritually) conscious retarded crippled , now to show an animals or an object as human can be funny is still a disease but one can enjoy little a bit of imagination , for example in Japan are doing this very often for every things , they make cartoon for showing product brand of factory to create a synthesis so the product brand is record nice immediate , that can be funny and useful for the market colorful is a minor form of disease of anthropomorphism , but to that about God to personalize God as human being and give appearance character behavior is the apex of the disease call anthropomorphism to give a face to a faceless reality quality of the universe , is a disease that make human being infantile retarded bigot , what to do this humanity is hypnotize with such a fairy tale so infantile is amazing how they believe blindly in such a lie absurd tale, out of date expire lies , you know a lie repeated at infinite nausea it become real, that is the neck of lie repeated at infinite it become real, but neither the less is a lie the truth is that is a lie , and this blind humanity it seems that is not interest on seen the truth they are at easy with lies and hypocrisy amazing , So anthropomorphism is the outcome, a disease, man anthropomorphic infantile, retarded, neurotic, in sort pathologic, neurotic because is neurons vector of emotion believe thought they function as a chain reaction into the brain, have being feed with lies unconscious staff of tradition of the past wrong information, conditioning of all kind and repression of all kind and sorry garbage of all kind, so neurotic. To me exist a faceless reality that is just a quality a creativity a pulsation of love intelligence, of the universe a fundamental law of the universe call it empty consciousness no adjectives no labels no contents , just an empty consciousness neutral to any interpretation of the little men, neutral to gender color race age , is just a quality a pulsation a creativity intrinsic to the universe that modern physic call it Boson x the quality that hold together keep together the all universe and confer Mass to particles atoms and matter , without the Boson x the all universe and forms of it disintegrate instant , is the quality pulsation that confer organism to the universe otherwise the universe would have being a sum of part mechanically assembled , but is not so is the Boson x or call it empty consciousness better that confer Mass and organism to the universal body , and the human being is an organic unity of it , through is inner being is link connected to this faceless reality quality pulsation called universal consciousness , his inner being is the bridge that connected him from the known to the unknown faceless reality quality pulsation creativity called empty consciousness . The path to it meditation , silence , love , this are the path to your inner being , and then you are link connected one with the empty consciousness fundamental law of the universe, a faceless reality beyond any interpretation of the little men retarded infantile neurotic in sort pathologic .Basic the path to your inner reality you close your eyes the present is the gate less gate to the vertical dimension of deeper and higher and you let the conscious energy flow within the so call dark matter or gravitational energy field will naturally bend the light consciousness deep within your inner being , you relax get centered into your inner being into your witness consciousness, at the deepest point of your inner being the pin drop silence is the gate less gate to your inner empty consciousness, universal consciousness the so called dark energy or expansive energy field will expand carry your light empty consciousness at the horizon of event of the universe expanding with the same passé of the universe on and on, and you in essence will be one intrinsic to the universal body is a break through divine sacred , your inner being at this stage is the whole universal body and your size is the size of the whole universe, is majestic divine sacred , and then on and on the so call dark flux or flowing energy field will flow move circulate your light consciousness on and on, in and out of the very source core of the universe , non being body or called it eternity itself , at this stage then nothingness emptiness is the gate less gate to non being body were time space forms duality all annihilate even consciousness annihilate or better twisted turn into unfocused awareness relation less , no more time or space or forms or duality consciousness as no more relation to be, so it annihilate into awareness unfocused relation less is just an I am ness, into non being body into the core and source of eternity, which is the core and source of the mystery of the universe and of life and death and of all duality , here you in essence are at home awake from unconscious sleep , resurrected , immortal the so called resurrection rediscover , and enlightened forever for eternity to come ,the real freedom unfold the real peace unfold the real silence full of subtle ecstasy unfold is a divine sacred bless full opening, into an open relativity endless begin less never born or die that the meaning of eternity , but is absolutely incorporeal is a conscious alchemy beyond above transcendental to whole and every things , That is it the truth , no face or figure of anything's is find, is an incorporeal dimension a conscious alchemy from unconscious to consciousness to awareness to non being body to the core and source of eternity begin less endless never born or die . That the true reality faceless beyond the disease of anthropomorphism, which to me is worst then cancer , because it create thousand and one conflict into this world today , all the out of date expire religion are at war with each other at the neck of each other for futile motivation, which are absolutely absurdity neurosis not exiting anywhere if in the imagination of the anthropomorphic little men , Is very important that the world get this insight that I am spelling here because the world is on the verge of a total destruction through nuclear war , A global suicide , for futile motivation , lies neurosis , difference of ideology and tradition , basically all tradition of the world have to be abolish band , starting from the tradition of war it was an ancient barbaric unconscious tradition of the world that they live interact through wars killing and destruction they were primitive barbaric unconscious infantile childish , they have no better way of interacting so they were fighting wars , but the world as evolve immense through meditation consciousness awareness intelligence , that this tradition of wars is totally out of date expire , the nation of the world today can interact maintain an equilibrium through intelligence consciousness awareness, love compassion peace , playfulness , the barbaric horrific unconscious tradition of wars as to be totally abolish band as a crime against humanity , enough of who has the most deadly weapons as the power of the world this is animal like the monkey gorilla act like this , all the most deadly weapons have to be abolish band deactivate as a crime against humanity , and all the conventional weapons too as a crime against humanity , no more building of weapons of any kind is abolish band as a crime against humanity , all the nation who as this deadly weapons from nuclear hydrogen bomb atomic bomb conventional weapons have to make table of peace dealing and deactivate destroy all of them , all the nation who hold such a deadly power have to come together and discuss how to deactivate destroy abolish all of this deadly device that are a danger for humanity not any more affordable , not one or two nations all of the nation who hold such a deadly power , are abolish band is a crime against humanity to hold them or to build new one , no more tradition of wars of interacting through wars , this was the way of the ancient barbarous unconscious humanity the way of the old little men , the new men meditative conscious aware intelligent interact through intelligence, love compassion awareness consciousness , and all military , weapons , wars , are abolish band as the worst tradition the ancient humanity as live to us as heritage, we don't want this heritage of wars military destruction killing deaths , finish abolish any military wars weapons existing weapons or builder of weapons are making a crime against humanity, that live conscious aware and in consciousness we are all one continent extension of one another just an oceanic consciousness , in the forms seven billions in consciousness just one oceanic consciousness and is a crime to kill or hurt an extension of your consciousness indirect you are hurting yourself your consciousness/ So is a crime and a crime against humanity , no more wars no more military no more weapons deadly and conventional , this are the first step towards civilization without this first steps civilization remain just a beautiful talking dream but never as exist civilization in this world , finish this absurd tradition of killing and destruction and deaths , the collective DNA of humanity demand this convergence of evolution immediate urgent now this mutation in better quality of living better shape of living in peace compassion love harmony, balance intelligence consciousness awareness is urgent now or never this steps forward toward civilization no this step no civilization .....Angelo Aulisa

The new book Mysticism & physic , Chapter 4 , Angelo Aulisa

Anthropomorphism part two

Hi friends, the subject of civilization and a world without weapons deadly like nuclear , hydrogen bomb atomic bomb , and deadly conventional weapons , and ashamed on you chemical weapons , is very important for the world today to understand how far in madness neurosis dementia senile the world as gone , the amazing thing is it seems that the world as no better things to do that thinking how to kill how to destroy how to make bloodshed , it seems that this humanity is blood testing , no sign of love compassion intelligence meditation , consciousness awareness celebration anywhere , is incredible how loveless as become this humanity , all the nations are bend in production building of deadly weapons that are meant to kill millions of people if delivered , is totally insane world mad world , the nation's interact between them through deadly weapons who as the worst deadly device weapons has more power , is not this terrorism , through fear scare the other nation , this is animal like the gorilla act like this in their tribe , the stronger gorilla is the head of the tribe , but for them is fine they are unconscious animals we can understand but how it come to this to human being this we cannot understand or maybe is the identification with the past ancient unconscious tradition of humanity , they were primitive barbaric unconscious not evolve and they have no better things to do they were infantile childish, with absurd ego trip power of conquering the all world , We have to get reed of this tradition of wars abolish band, the new man as evolve in intelligence meditation consciousness awareness, we have thousand and one subject to evolve better things to do and to think about , love compassion , meditation consciousness awareness, celebration of all kind , painting , sculpting , wisdom knowledge , the discovering of the universe and all of is intrinsic law , a great mystery , and the discovery of the inner nature of men the inner science mysticism etc.. this tradition of wars interacting between nations through deadly weapons keeping the equilibrium through weapons is out of date expire finish, to abolish is a crime against humanity and the intelligence of human being a big crime , But neither the less all the politicians are dealing on sale of the most deadly weapons , is their work they are broker of weapons , graveyard diggers , this politicians of today they have dig a big graveyard for the all world with the selling and dealing of all this deadly weapons , they are nuclear weapons for destroy 15000 time the all world that means that every men can be killed 15000 time but one time is enough and is dead , ashamed on this politicians , by the way I want tell them that they are other deal for making money if you like so much money, many other items are there to deal with that are beneficial for humanity starting with equipment to reduce pollution for example , I want ask them the politicians how it come to this point of neurosis madness you have no better things to do or to think about, like love celebration playfulness arts of all kind sports of all kind , please stop abolish immediate urgent this tradition of wars , weapons building military , killing destruction deaths and bloodshed , this is unconscious barbarous primitive infantile childish , and dangerous to the max , this is a crime against humanity , against civilization which has never exist in the world up to today , the first step toward civilization is the abolish band deactivation of all the weapons existing today, deadly conventional chemical all of them , all the nations who hold this deadly power device of death have to come together in table of peace and talk how to destroy deactivate all of this deadly weapons, not one or two nations but all of them together and simultaneous , is urgent immediate deactivation of all weapons this folly of the past ancient tradition of humanity of wars is abolish band , no building of new one the proliferation of any nuclear or conventional weapons is a crime against humanity, and the one who those this proliferation will be brought in international court and face justice , The new men the new world the new U N , want live in peace harmony love consciousness awareness, civilization intelligence meditation , and is a priority the abolish and band of the all weapons existing one and stop band and closed of factory industry that produce such a deadly device of death and destruction , we change the way those factory industry will produce equipment to reduce pollution into the world in accord with science and physic . No condemnation but just a total change of gestalt of focused toward peace civilization intelligence love , celebration playfulness , arts of all kind painting , music singing dancing , sports etc.... a new humanity a new era of celebration peace love meditation, consciousness awareness balance harmony etc...Basically is the dropping of all tradition of the past ancient humanity 99 per cent of those were just garbage believe , a nightmare of the past unconscious humanity a bunch of neurosis , of the past primitive ignorant unconscious humanity , starting from the tradition of wars military weapons , basically a world without wars, military , weapons, for few dozens of years we will live only the police station into the world that they will have the mission to monitored the transformation of the world towards a conscious living an aware living a loving compassion living in harmony peace and balance only the police station worldwide are enough , and after 100 years also the police station will be not needed because we will have a conscious aware humanity , basically all source of crime violence is the unconscious staff of humanity wen the unconscious is no more crime disappear violence disappear . This are the first step towards civilization the abolish and band of all weapons and military and weapons , this will give great resource to humanity economic resource to evolve in other direction beneficial to humanity and is urgent needed to eliminate starvation poverty misery , without this first step we cannot talk of civilization , for example America is not a civilize country were u wear a weapons to go out in the community that means that there is no common sense , people are afraid of each other so they wear a weapons to go out in the street this not civilization , but the law of the jungle animal like , worst animal don't wear a weapons , is insane , and America population is a population of obesity were people eat night and day nonstop because basically they don't love they miss love, so they eat all the time a loveless country , this is not civilize , America is not a civilize place but very barbarous unconscious country and they produce weapons of all kind deadly device to kill people is full of factory industry that produce weapons of all kind, and they sale all over the world creating proliferation of wars deaths and destruction and bloodshed this as to change dear Americans, I say this with love and not condemnation , please change the way you are interacting in your country and with the other nation too , enough is enough , they are other item to deal , beneficial for humanity if you like money so much , but is not only Americans all over the world same story China , Russia , Europe , etc...no more production of deadly weapons conventional weapons chemical weapons is a crime against humanity, and a crime to kill an extension of your own consciousness every living being is sacred holy divine, you change production in item that are beneficial for humanity, are u crazy mad enough, turn your attention to life to love to celebration, comedy are more beautiful than tragedy, so that is it if u want a better world civilize I have indicate the first step to do absolutely urgent necessary now. Then I would like to change subject , wen I say that entanglement is a phenomenon that is beyond life and death and is connected with past life, that u can uncover past life , is true 100 per cent , but to be more accurate the entanglement of two being transcend life and death it remain forever , but when one of the being entangled leave the body it goes into a journey of reincarnation and within days months it get a new body the entanglement is on in this world objective, you may meet again the soul that you were entangled in this objective world , is not and is useless that you try to meet the soul spiritually that soul is already born again into a new body, in days months so if you try to meet again spiritually body less is not the case , the soul that goes for a journey of reincarnation get a new body very soon , the entanglement process happen objective in this world you may meet again the soul that you was entangled with , but here in this world objective , and many of the people you have relation are those soul that you were entangled in past life, you need sensitiveness intuition to record nice and feel, if you try to recall the soul spiritually body less is a laughing stock, is not anymore in the journey of reincarnate most probable as already reincarnate it need days months and is already reborn , there is a mysterious attraction here in this objective world that you meet again the soul that you were entangled in past life , and many time is the case of all your friends parents etc... but here in this objective world you need intuition sensitiveness and a no mind approach, a meditative approach and you may record nice the soul that you were entangled , this experience give you a confirmation of life that is eternal and death do not exist which is great you learn that death is the greatest fiction of humanity, the forms dies but consciousness is eternal deathless until a final annihilation into the core and source of the mystery of the universe and of life and death and of all duality into the law of eternity , then you will have no more reincarnation but for eternity to come u will be in essence data intelligence into the universal body , of course body less, free of the cage of the body expand into the universal body flowing through consciousness universally forever for eternity to come, that is the resurrection a conscious alchemy not material not physical not gross , you will dance body less into the universe in essence you like it ......Angelo Aulisa

The new book Mysticism & Physics Chapter 5 , Angelo Aulisa

Consciousness & BBC

Hi friends , just now few minutes ago the give a news on BBC , that a patient that was in a coma for 15 years in a vegetable state , the doctor have being able to restore consciousness into the brain of the patient through an operation , a surgery , they have implant some tissue that were missing . Now if this was true was the greatest news ever since ten thousand years , be very careful in use terms not accurate , Consciousness is not such a thing , first of all is not a thing consciousness, is not material is incorporeal not physical not gross , you cannot grasp it and put from one plays to another place , if it was so then we can make the all world conscious through an small surgery operation , wow , was the great discover ever done ever , but unfortunate is not like that , say that you have restore the organism body of the patient all the chemical reaction of the body organs , and the mechanism is going to function again properly , and most probable wen the mechanism of the body is restore also the neurons into the brain will start function again and create chain reaction into the brain and they will function the neurons as vector of thought , emotion , believe , and the patient will have all the organism mechanism of the body working function properly , but as a machine mechanically through the chemical reaction of the organs of the body, and that is very beautiful a great happening into a patient that was 15 years in coma my gosh I don't want say that this is not a beautiful happening , great science is doing miracle in this day great . But don't called it the restore of consciousness , yes consciousness is everywhere and nowhere in particular expand into the universe a fundamental law of the universe is just a quality a pulsation of love intelligence , a creativity intrinsic to the universe but incorporeal sacred divine , you cannot grasp it with your hand and put from one place to another , or buy it or sale it , is a conscious alchemy not gross not material not physical , physic took 40 years of experimentation to have just a glimpses of the boson x they have discover they have seen the quality called consciousness into a laboratory but just for a split second and you cannot grasp it or isolate it they see the quality called Boson x , and they see that is the quality that confer Mass to particles atoms matter and they see that is the quality that hold together the universe and without the Boson the universe disintegrate instant and all forms of the universe , because when you take away the quality that confer Mass to the particles atoms matter is obvious that the universe and form of it disintegrate , but they cannot isolate and grasp it I guess , and make injection of Boson x . No Consciousness remain a mysterious phenomenon happening sacred divine , the path to consciousness silence , love meditation , laughter , dancing , those path are called path because they take you in essence beyond the mind , into a state of no mind , no thought no emotion no unconscious staff wen the mind is no more and only a pin drop silence is that is the gate less gate to consciousness when you are totally centered relax into your inner being at the deepest point of your inner being a pin drop silence , and your inner being is the bridge the link to consciousness universal consciousness you in essence become one in mystical union with consciousness intrinsic to consciousness , and consciousness do not reside into the brain is a fundamental law of the universe expand everywhere and nowhere in particular into the universe you can be link connected to consciousness through your inner being bridge link to it to consciousness , then your body from a mechanism that was become an organism an organic unity with what, with the whole universe with the organism of the universe , before that this alchemy happen you are just a mechanism unconscious even if you are healthy , you are a mechanism of chemical reaction unconscious and the chemical reaction the energy that fuel it are the aliment food that you eat , but unconscious all the same , the unconscious you built up from the day that you born up to today present life is a huge immense accumulation of conditioning repression , wrong information projected on you by society system religions family , all outer sources , basically is your neurons have being feed with all of this staff of collective unconscious and they function into your brain as chain reaction vector of thought believe emotion that have being feed by the particular society that you live in , for example an Indus as a particular conditioning a Christian another a Muslim another a Buddhist another a Jainism another and they are hundreds of society into the world all different from one another and all the people neurons have being feed with different information staff , and they function accord to what as being projected in that is all unconscious staff , basically 99.9 per cent of humanity is unconscious even if their body function properly healthy are unconscious , that is why I insist always that the neurons should be left clean free without any conditioning repression unconscious staff , because a free neurons clean become vector of consciousness it flows through the neurons then, and consciousness intrinsic to it as great quality all the essence dharma are included in it is the source of love of intelligence, of harmony peace joy , creativity compassion etc.. so your neurons will be vector of all this divine sacred quality but they should be at priory empty clean free , if the damage of society religions system family conditioning as being done already you are unconscious that what in the end the unconscious is your neurons full of garbage of conditioning repression of society system religions culture philosophy etc…and 90 per cent of cases the unconscious the hypnosis as gone so deep that is irreversible you cannot clean evaporate anymore wash anymore , and you live a life unconscious at the mercy of your dirty stinking neurons full of garbage , they have to be clean wash , but many cases are irreversible , yes you need a brain wash because is dirty since dozens and dozens of years you have not clean wash it and it stink is dirty , laugh please yes a brain wash is necessary so that neurons turn clean free again and become vector of consciousness , but is a difficult task to clean the unconscious to evaporate the unconscious , through meditation and hundreds of method of therapy they can be clean again and the neurons turn again free clean of conditioning unconscious staff wrong information , and you can see again the life through your true nature original . So don't say that the patient as being restore consciousness is a big lie , and my gosh in a particular place into the brain , look the link bridge to consciousness is your inner being , were is your inner being your soul into the Hara the link with the body , and it encompasses spherical your organic unity at certain distance all around , that is why you can see your body your thought your emotion your senses , at a certain distance, and your inner being is not situated into the brain , is a mystical alchemy divine sacred were is required to be centered into your inner being , rediscover of your inner being realize your inner being and be centered into it, then you can be linked connected to consciousness in mystical union one wit it intrinsic with it , is a long way a long path , if that was possible with a surgery was the greatest discovery ever made , is a surgery but mystical, of your unconscious of your neurons but incorporeal, and there is no guaranty of success you may rediscover your true nature of your inner being you may not go through no guaranty, it depend or your intensity afford that u put into the experience it depend in your decisive approach , many time they use wrong terms don't call it consciousness, or we raise awareness no awareness cannot be rise we raise consciousness about a subject of climate change for example, but you cannot raise awareness, because awareness is relation less unfocused an I am ness, already the case there …..Ok sorry for the accuracy but is important I love BBC news ok Angelo Aulisa

Published on September 26, 2017 14:37 • 67 views

The new book Mysticism & Physics Chapter 6 Angelo Aulisa

Consciousness & BBC part two

Hi friends, so consciousness is above beyond transcendental sacred divine , incorporeal , cannot be taken from one place and put into another place , cannot be sale or buy , basic is a fundamental law of the universe everywhere and nowhere in particular expand into the universe ,a quality a pulsation of love intelligence a creativity intrinsic to the universe , no label or adjective or contents can define it consciousness , is infinite it goes as far the event of horizon of the universe goes , then it annihilate into non being body , which is not a body is incorporeal non being body , no space no time no forms no duality are find into non being body but just awareness relation less unfocused is just an I am ness awareness a great relaxation , silence a subtle ecstasy is intrinsic to the awareness ultimate essence of the law of eternity and awareness too is everywhere and nowhere in particular expand into emptiness nothingness into the law of eternity , incorporeal , beyond above transcendental dimension less dimension . What I care to say now to be more accurate is that your inner being encompasses spherical all around your body within and without at a certain distance, that is why you can see your body your thought your emotion your senses, but the link were is connected with your body is the Hara Chakra , and the way that consciousness flow circulate moves into your body is through the all seven chakra , when the conscious energy is release through pleasure it move down wards into the sex center and momentary pleasure is the outcome , when it move upwards through the all channel of the seven chakra , from the Hara to the navel to hearth to the expression to the third eye to the seven center the lotus situate in the back words of your head above the small brain , and further into a mystical union with the universal consciousness, a mystical union happen with the organism of the universe you in essence become one intrinsic to consciousness, is a break through sacred divine a conscious alchemy sacred divine your size change you become big as the universe is big , is a majestic divine sacred break through a mystical union with the organism of the universe, and then you move circulate flow your light consciousness in a conscious alchemy , spiritual to talk in old ancient terms . The function of the mechanism of the body organs and brain can be restore through science, medical, the doctor can do that, is great they can restore health, but consciousness is beyond above transcendental to the mechanism of the body, the organs the function of brain can be stimulate in thousands of way and you can restore the mechanism of the body, for example Delgado device a chip that implanted into the brain of a rat can stimulate the center of orgasm, and with a remote control. he just press a button and give an orgasm to the rat , and he those that for hundreds of time and the rat get hundreds of orgasm until the rat die of orgasm , it can be done into human brain too , it will be very funny you just press a button and you get an orgasm as many time you want , even until death of orgasm laugh please smile , the mechanism of the body can be stimulate in thousands of way , but consciousness is above beyond transcendental to whole and everything incorporeal divine sacred , to be link connected love silence meditation, laughter dancing singing painting basic all method that bring you into a stage of mindless, no mind beyond the mind , were no thought no emotion no senses are left but a mindless pin drop silence, and your witness consciousness is find your true nature is rediscover, at the deepest point of your inner being you get linked connected with universal consciousness in a mystical union one intrinsic to it, your inner being is the bridge the link from the known to the unknown infinite consciousness , that is it , by love all of you ...Angelo Aulisa ...

Published on September 27, 2017 00:23 • 53 views

The new book Mysticism & Physics Chapter 7 , Angelo Aulisa

The history of the universe comment

Hi friends , I would like to comment on the history the universe in two hours , a documentary that I have seen recently , first of all the documentary history is all about the rudimentary objective part of the universe what unfold after the big bang into the objective universe , same is all the physic and science explanation of the universe, objective the material part of the universe , on this regard I would like to say that even on the rudimentary objective explanation of the universe much is missing , if you want an explanation analysis that pierce the objective analysis please read Hi light of an unknown author called Angelo Aulisa and the Quantum leap of the same author and many other books of the same author Angelo Aulisa , anyway I would like to added few particular , after the big bang we have two billions years of dark age called in physic , in the dark age of the universe after the big bang we have a big huge cloud of dust gas debris scattered everywhere and nowhere in particular in the universe, it was a huge cloud bigger than a galaxy even more , that for two billions years the dark age of the universe was flowing all around the space , is important to came back for a while at the big bang event , with the big bang all the law of the universe came to be , I want make an accuracy , the inflection on even today expanding the universe at first, but basically all the law of the universe known and unknown came to be simultaneous to the big bang starting with inflection but to be more accurate the really first law of the universe that came to be simultaneous to the event of big bang is the Boson x , that confer Mass matter to the particles atoms matter , in fact the huge cloud of dust gas debris the dark age of the universe is create by the Boson x that give Mass matter to gas dust debris , physicist tend to forget to mansion this majestic law of the universe without the Boson x they would have being no Mass no matter of the huge cloud of dust gas debris , and basically in mystic terms simultaneous to the event of big bang consciousness came to be in relation to the subject of the universe itself , now called it Boson x or called it consciousness the intrinsic fabric of life of the universe is the same quality two name for the same quality called Boson x that confer Mass to particles matter atoms , so I would say after the big bang event all the intrinsic law of the universe came to be simultaneous , starting from consciousness or Boson x that in relation of the subject of the universe came to be and confer Mass to the matter of the universe the huge clouds of matter of dust gas debris called the dark age of the universe that last two billions years , and simultaneous the law of inflection that expand the universe even today, and simultaneous the law of dark matter or gravitational energy field that for two billions years to sculpt assemble the early universe, the dark matter is the great painter the sculptor of the universe due is gravitational energy field expand everywhere and nowhere in particular into the universe , the universe that you see is the work the function of the gravitational energy field called dark matter and it came to be simultaneous to the big bang as all the law of the universe ,Then the dark energy or expansive energy field of the universe that is expanding tearing apart the universe every slit second and tearing apart all galaxies planets stars, is an expansive energy field that pull upwards the gravitational energy field so called dark matter is pulling Inwards down wards and this two law of the universe that are one is 24 per cent of the universe that dark matter, and the other is 70 per cent of the universe the dark energy are a perfect polarity of complementary opposite, but complementary that balance each other due the fundamental law of the universe called the law of dialectic , not very perfect because the share of those law is very different the dark energy is much more powerful is 70 per cent and for this in the long run is going to win the race with the dark matter and will disintegrate literally the universe after billions of years, the universe will be a cold frozen void nothingness emptiness the dark energy will win the race and fulfill is task in disintegrate the universe , but for a while the law of dialectic of opposite complementary will hold on the universe in a balance harmony as we see , but then the 24 per cent of dark matter will lose the challenge race , and the dark energy will take on and make of the universe a cold frozen void nothingness emptiness , I say the dark energy will fulfill is task because the universe which we see in matter is only 5 per cent of the totality of the universe, and the universe is a cycle many universe have being before our universe and many universe will be after our universe is like a cycle of season, due the fundamental law of the universe called the law of dialectic , when our universe will be just a cold frozen void a nothingness an emptiness will create the polarity of infinitesimal micro, singularity of the atom which is the infinitesimal, smaller singularity that at is turn will create the complementary of whole fullness, macro, the friction of thesis the singularity of the atoms and antithesis of whole fullness macro, will create a new event of big bang, a synthesis of big bang at infinite , many cycle of the universe have already being many season of universes have already being but the law of dialectic of opposite complementary will recreate a new universe at infinite endless , eternally . in fact the event of big bang take place always and always into a canvas of eternity bigger and vaster then the universe itself and begin less endless never born or die, that the meaning of eternity is a canvas a dimension less dimension above beyond transcendental to the universe, and core and source of the mystery of the universe and of life and death and of all duality ,in fact time space forms duality totally annihilate into the law of eternity , is an endless journey into eternity death do not exist either for the universe either for human forms , but the reality is immortality and resurrection , the resurrection is a conscious alchemy not physical not gross not material but a conscious alchemy, from unconscious to consciousness to awareness into non being or law of eternity . Then the law of Dark flux or flowing energy field came to be but I remind you all the law came to be simultaneous to the big bang so the flowing energy field called dark flux that move flows irrational planets stars galaxies at the event of horizon of the universe, and make them disappear into non being or the law of eternity, and reappear into the universe again and again, in and out of the source and core of eternity in a cosmic play , very important the dark flux , all of this law play their role and influence the universe every split second, and same they do into the human forms into the organic unity of body and into the organism of the universe, they are expand everywhere and nowhere in particular into the universe and in and out of our body too , our body is keep together by the dark matter gravitational energy field, for example ,, so all the forms of the universe trees rocks, animals water stars planets galaxies , all the law of the universe play their role game into our planet and life every split second , for example I say this many time the gravitational energy field called dark matter bend the light consciousness deep into our inner being , as gravity bend light time and space and create gravitational waves this my discovery stolen by the scientist two years ago , first I publish the introduction of Le siècle the lumiere were I report this 2 of February and BBC after ten days give the news of the discovery of gravitational waves ha ha ha , and same the dark energy or expansive energy field pull upwards the light consciousness expand it at the event of horizon of the universe on and on, is the law of grace , and same the dark flux or flowing energy field is flowing moving circulate the light consciousness in and out of the core and source of the mystery of the universe into the law of eternity or non being body , in a festival of light consciousness is moving flowing the light consciousness in and out of your organic unity in a circle infinite with the universal body, and the core and source of eternity non being body eternally , all the intrinsic law of the universe play their role game into planet earth and into our organic unity eternally , now come tomorrow in BBC and say that Oxford discover this and you take the award of the discover , the world is a fucking sucker of other people intelligence discovery and they claim the discovery as their own ashamed less . For example a galaxy is keep together by 3 law basically , the major player is the dark matter or gravitational energy field that assembled together the galaxy with is gravitation pull expand everywhere and nowhere in particular it assembled dust gas debris into stars planets and finally it pull together the galaxy it assembled it, is well known , but is not all another major play is the electromagnetism , and another major play is the black hole with is pull it make rotate planets stars around itself , and if they fall into horizon of event of the singularity of a black hole, they squash disintegrate but if they don't fall in the black hole contribute to keep them rotate into the galaxy they belong, because every galaxy has black hole they are short cut warm hole to the core and source of eternity non being body, because into non being no space no time no forms no duality is find same as the singularity of a black hole , and a new discovery done by the Israeli scientist into the singularity of a black hole the data survive the intelligence survive they have isolate a particles that is the data the intelligence that survive into the singularity of a black hole I confirm it awareness survive is the ultimate essence of non being body or eternity called as you like, it survive always and always and it preserve essence fragrance intelligence because incorporeal, awareness is where the resurrection immortality take place and the data of intelligence refine by you in your life survive the essence survive the fragrance of your consciousness survive eternally that is what actually is the resurrection, now the Israeli scientist have isolated a particles that preserve the data intelligence even after the disintegration into the singularity of a black hole, this is divine sacred and true , Last I want say about the history of the universe objective everything well but I do not understand when it come that we master fire earth , and we create iron etc. how is possible that you emphasize the creation of weapons so much , ok in the primitive men unconscious barbarous he use those instrument weapons to hunt and let say ok in the primitive men that they fight wars to conquer the world they were primitive barbarous infantile childish unconscious they have no better things to do , but when it come to our age you still go on to emphasize the weapons and wars as a great discovery this is absurd , you know this so called civilization as fight more than ten thousand wars and kill and destroy millions and millions of people , causes destruction horror bloodshed massacred incredible , at a certain point the evolution of humanity as being perverted people have gone crazy by keeping on the tradition of wars as a way to interact between nation , basic is the identification with tradition all of them a bunch of neurosis a night mare of the past humanity ancient expire out of date unconscious totally, I will call it the dark age of unconscious of the ancient primitive barbarous humanity , Christians have a terms the dark night of the soul that are those traditions of weapons military and wars , this perversion of keeping on the traditions of wars military weapons as to ceases stop abolish band urgent now , 99 per cent of tradition have to be abolish they are all of them a night mare of the past ancient primitive humanity unconscious , a bunch of neurosis but at first wars military weapons, the first step towards civilization this attitude of considering wars something to remember preserve is out of date expire, they were just tragedy failure of intelligence neurosis , the new humanity interact between nation through intelligence consciousness awareness, peace love harmony balance love compassion , playfulness , enough of tragedy bloodshed massacre this is insane a dementia senile of humanity, a neurosis of humanity that as to ceases , this is dementia altzaimer losing of memory of location of where you are is a violent form of altzaimer , because you do not know where you are who you are, where you come from, where you are going, fear of life and death, the huge universe scare you want kill everyone in a global suicide , stop abolish finish of building weapons the factory who build weapons are doing a crime against humanity, they should be closed now, and interacting through nation with intelligence consciousness awareness ,no more military no more wars this are the first step towards civilization urgency needed, a changing of attitude of all the nations make table of peace deal, and destroy deactivate all the weapons deadly conventional chemical right now , if we want save the planet , this planet the lotus paradise and this body the very consciousness awareness, then civilization will flower never before this first step, they will be only talking dreaming but civilization as never being in this planet never ....Angelo Aulisa

Published on October 01, 2017 05:50 • 32 views

The new book Mysticism & Physics   Chapter 8   Angelo Aulisa

The history of the universe comment part two

Hi friends , the history of the universe in two hours part two , and so civilization in the last ten thousand years as never being , they have being beautiful talking dreaming but not civilization , because the first requirement principle of no weapons no wars no military as never being fulfill , this humanity in the last ten thousand years as fight ten thousand wars creating the worst bloodshed tragedy and millions and millions of deaths , destroying entire city population killing millions a bloodshed endless horrific , even today politicians broker of weapons dealing and graveyard diggers , they have built dig a big grave yard for the all world , they are nuclear hydrogen atomic weapons to destroy 20000 time the world , each of you can be killed 20000 time and the worst part is that nuclear weapons if delivered they vaporize the location were have being delivered and they create contamination radiation that will go on for hundreds of years , the end of the green planet and millions of deaths , dangerous like hell , but neither the less politicians goes around the world, on the surface they talk of peace but in their own nation they prepare for war build pile up big stock of nuclear weapons and hydrogen atomic bombs , they are unconscious asleep this politicians they do not know the hell and dangerous that are creating doing , Jesus was right before die he say forgive those because they do not know what they are doing, he knows they were unconscious Jesus knows already the unconscious , But is almost impossible to forgive such a criminal that they perpetuate crime against humanity in day after day second after second , building weapons deadly, selling weapons all over the world creating proliferation of bloodshed wars destruction tragedy , killing of children woman and civilize population mercy less, they are monster psychopathic . Friends until the world understand that we are 7 billion on the surface on the forms, but in consciousness within we are all one continent extension of one another sacred divine we are one organism of living being extension of one another, and killing hurt an extension of your own consciousness, being, is a crime and a crime against humanity , All living being are sacred divine , and this is the begin of civilization , so weapons builder are doing a crime against humanity day in day out they have to be stop abolish band, any factory industry producing weapons as to be shut down closed, they make crime against humanity and change production in beneficial item for humanity the factory who do not comply the with the new UN rule will be brought in front of criminal court and face justice , Consequence all military are abolish band we do not need them anymore they have to change work they should do things that help humanity recovery of calamity helping the civilization to grow without weapons , and consequence also wars will disappear stop abolish band , The problem is that all the nation together have to make table of peace discussing talking the subject of deactivation of deadly weapons, like nuclear hydrogen atomic how to destroy them deactivate them but all the nation together simultaneous, otherwise one can say u begin the process of deactivation of weapons an another nation will say you begin to destroy your weapons, and they will be no trust , but all together simultaneous the peace deal can be done , enough of mechanic unconscious that cannot be stop because the deal is so huge and so mechanically all politicians go on doing it ashamed less , finish of this attitude, make a meeting all together and change this unconscious attitude all together, and make table of peace rising consciousness that a total destruction is at the gate , what you want live in peace love compassion aware conscious, or total destruction I think intelligence will prevail , and the abolish deactivation of all weapons and stop abolish band of building new one, is a crime against humanity and human being enough of killing destruction bloodshed tragedy, I think common sense will prevail and intelligence will prevail, then only the first step towards civilization before that we are living an unconscious barbaric age the dark age of the unconscious . Then I would like to added to the history of the universe a few things , the first is that after the big bang event the first law that unfold is the begin of time and space , I forget yesterday this basic law each universe at the big bang time event, unfold a new begin of time and space and all the other law of the universe known and unknown unfold simultaneous as I say yesterday , And there is an interval from the ending and disintegration of one universe to the begin of another due the law of dialectic, which trigger a new universe endless eternally as I say yesterday , when the universe is at a stage of cold frozen void emptiness nothingness due the disintegration creating by the dark energy , there is an interval were the singularity of infinitesimal of the atoms is getting momentum probable of billions of years, that I would like to call this interval from one universe to another the sleeping age of the universe, where there is nothingness emptiness cold frozen void darkness all around , Wen the singularity of infinitesimal of the atoms get ready is at the apex of infinitesimal micro, creating a perfect thesis for the unfold of the antithesis of whole fullness macro, friction of those and synthesis of a new big bang and a new universe is taking place shape , I think by joking that a new universe is the only hope because this humanity civilization as fail totally in all the field of life in understand life and as create a hell on earth , as fail totally in the inner reality of men as fail totally in the objective life of men as create a real hell on earth that the only hope is to wait a new universe to unfold . This humanity is a great failure I am very upset for this failure, we would have being if the enlightened one have being heard the greatest civilization of the all cycle of universes ever happen , the apex , because many civilization have being during the many cycles of universes but none of them have reach even closed in what I am saying to you , it would have being the triumph of our DNA which aspire for the ultimate convergence of evolution quantum leap into immortality resurrection then the DNA as accomplish is task of convergence of evolution and is enlightened in a relaxation , for example yesterday I talk to you that Israeli scientist have isolated a particles that survive the squash disintegration at the horizon of event of the singularity of a black hole, is true I have read it somewhere , the data intelligence survive because are incorporeal cannot be squash disintegrate by the black hole , and is an inference of all physicist that the data survive the intelligence survive, but apart from this I tell you always that awareness relation less unfocused is the ultimate essence of non being body, were the intelligence the essence the fragrance the data of all enlightened one survive eternally, awareness is where the resurrection take place reside , is like you added a lamp in a room where they are already thousands of lamp, it only intensify the light that already above into awareness without disturbing other lamp light it will create more intense light , when we say light consciousness or enlightened is because is really like that consciousness awareness are intense light transparency whiteness in that light transparency whiteness are intrinsic incorporeal all the essence data intelligence fragrance of all the enlightened one ever happen, and that have had a resurrection and of all the enlightened one that will happen and of all the enlightened one happening right now ,into the singularity of an atoms protons neutrons and electrons are like past present and future, into awareness they unfold simultaneous , too difficult for you to understand I am sorry get up date through my books the quantum leap , hi light , and 12 books more are there , we would have being the greatest civilization the apex of civilization of all the cycles of universes ever happen a triumph of intelligence , but the old little men the identification with tradition of the past dead old and out of date, and the expire out of date religions are creating thousand and one hindrance for civilization to happen, and for the evolution of our DNA and intelligence to happen, they should resign all of them the out date religions they have done their job work is time for them to resign and let humanity evolve in intelligence freedom civilization consciousness awareness , physic science have out date expire all the religion totally demystify all of them, they are a laughing stock of absurd superstition dead tradition that is leading the world to a total destruction, enough please resign peacefully all of the out of date expire religions now without hunger, I tell you this out of love and understand of the reality of our contemporary age 2017 get update in consciousness awareness, here now flowing in consciousness moment by moment into the present the only real tense ok love all of you Angelo Aulisa

Published on October 02, 2017 05:01 • 28 views

The new book Mysticism & Physics Chapter 9 Angelo Aulisa

Unconscious & conscious
Hi friends , there is no God and no evil this are childish superstition , God is anthropomorphism a disease of personalize God give a face to a faceless reality, and evil an invention of the priest to scare retarded people , basic all old out of date religion they play on your fear they scare you of hell fire for eternity and they give you hope of an not existing paradise , neither hell exist neither paradise exist is just a game to keep hold of you all so they dominate you, the priest the religion and you fall on the vicious game as puppet ,. What people call evil is nothing else that your unconscious , the unconscious is the basement of your organic unity, you go on throwing into the basement of your unconscious all wounds trauma conditioning repression that happen during the course of your life , whatever hindrance your freedom your spontaneity your true nature your naturalness you throw it into the basement of your unconscious , the unconscious start build up since you born , your neurons into your brain chain reaction of thought emotion believe and vector of those thought believe emotion start to be feed with wrong information conditioning repression, unconscious collective staff, mode of behavior of society family religions , is a projection into your neurons of all outer sources artificial , they are hundreds of society different from one another , dozens of religions different from one another and everybody get a different in print of neurons , and that is where the unconscious start built up day in day out every split second , your true nature your freedom your spontaneity is totally neglected , your inner being totally avoid , so the organic unity of human being as a safety measure called unconscious which are the basement of your organic unity and men goes on throwing into it all conditioning wounds trauma repression , since you born the unconscious start built up and it goes on up to the very present life is a huge accumulation of conditioning , the unconscious is nine tent of your life and the conscious that you show day by day is one tent like an iceberg that you see the outer part on the ocean but under the ocean is nine tent of the iceberg , same is the unconscious of a human being , it condition all of your life , apparent you may show your one tent conscious but the unconscious is always there getting momentum to assert itself , any situation any controversy and the unconscious take over bubble up with revenge aggression, violent and you cannot do anything is to huge nine tent it wash away the one tent conscious in a split second , and the unconscious is wild violent it knows only need ,. Now as I say before one goes accumulate all of his life into the unconscious trauma wounds repression conditioning, and the unconscious get saturated it stem up , and any controversy discussion that don't feet the projection that as being done into the unconscious and is ready to assert itself and jump on you violent aggressive , in extreme cases a pent up happen were the unconscious stem up all folly madness that as accumulate and then is very dangerous because one is in a rage blind to any reason rationality , the unconscious is absolutely irrational wild violent aggressive wild, in extreme cases dangerous . And so one in is day to day life he may never show any sign of madness because the one tent of conscious is on but any situation may trigger the nine tent of unconscious repressed underground , this extreme situation of madness violence aggression are called pent up of unconscious the unconscious stem up like a cocking pot , and make disaster horrible tragedy ,there is no evil which tempt you to do evil this is an absurd superstition it do not exist , what really exist is your unconscious saturated to the max , in fact today days 99 per cent of humanity live unconscious totally unconscious asleep , and everyone is a potential monster it can explode anytime with thousands of different neurosis , it depend what you have repress in your life the priest explode into pedophile for example but hundreds of potential disease neurosis are there into the unconscious is a junk yard of neurosis . That is why they are psychologist psychoanalysis to cure heal the extreme cases to heal the extreme cases , but they can do very little because as soon the session of psychoanalysis is over the next second the unconscious goes on accumulating again dust into the unconscious is a partial treatment that make people normally heal , they are treatment more complete therapy groups , method of healing , like dynamic meditation were you throw the folly at the wind in five stages very effective because it focus also on meditation it make you centered into your inner being, so any dust that you may collect again you watch at it from your witness consciousness from a certain distance, because you are centered into your witness consciousness, gibberish that means throwing at the winds all language known and unknown is a cleaning of your brain neurons , and many other method like the mystic roses one week silent one week laughing and one week crying done with therapist that help you to focus on meditation on centering you into your inner being, and witness consciousness , basic any healing of the unconscious that do not imply meditation and centering into your inner being, and witness consciousness is partial the method that imply meditation are total and a complete healing happen . Basic 99.9 per cent of humanity live unconscious and

100 per cent of society live into the collective unconscious , that is why is very dangerous to allowed selling of weapons because any moment any controversy any dialogue wrong interpretation and the unconscious is there bubble up and create disasters , a society that allowed sale of weapons in such a state of dark unconscious it may have dramatic consequence, for the innocent people who live in that society , there is no evil that tempt you to do anything it do not exist but certain exist an unconscious individual and collective that stem up again and again and create disasters calamity for innocent people who live in that society , what retarded people call evil is nothing else but the unconscious repressed conditioning, that make you do thing not even imagine , So my suggestion is to abolish band shut down the factory industry that produce deadly device that kill other human being is a crime against humanity to do such a production, they should change item of production make item beneficial for humanity , and for the next hundreds of years band stop not allowed any selling of weapons anywhere into any nations because this humanity is totally unconscious and anytime a pent up of unconscious may happen and innocent people die pay an heavy price with their life , just to make the lobby of weapons earning more and more money this lobby of weapons selling are monster psychopathic that they see only money and money, at the cost of human life of innocent people we should stop them band them abolish them right now and sorry fuck them up now , Either the change there production then fine or face international court for crime against humanity , enough is enough we live in new era were humanity is struggling to become conscious aware , and beyond into consciousness we are all one oceanic consciousness extension of one another into the forms seven billions but into consciousness one continent of oceanic consciousness , and every living being is sacred divine holy , and is a crime to hurt kill interfere in a living being life , so the lobby factory that built weapons are making perpetuating crime against humanity every split second , I say this thing out of love not with hunger to the lobby of weapons you have already make allot of money please give up the production of deadly device ,change production into item beneficial for humanity, equipment to reduce pollution for example have mercy compassion resign abolish such a deadly device production, do as Donald Trump those he sale cafe ice cream hot dog cake chips innocent things and he has make allot of money, you change production of item and is done the first step towards civilization is done , we live into a new era of meditation consciousness awareness, peace love compassion playfulness harmony, balance intelligence, the new men will never accept the tradition old and out of date of wars weapons disasters, bloodshed horrible massacre, is over , have a quantum leap convergence of evolution into consciousness awareness and change production make item beneficial for humanity and you are all welcome to join the new dawn of civilization consciousness label less adjective less content less, of intelligence of love of silence , this planet the lotus paradise this body the very consciousness awareness that is it .. Angelo Aulisa....

The new book Mysticism & Physics Chapter 10 Angelo Aulisa

Unconscious & conscious part two
Hi friends , first of all I would like to know why all media news are afraid to mansion the terms unconscious , in the last century giant of intelligence like Freud , Carl Gustav Jung , Adler, and many other psychologist bring to light from the ignorance of humanity a tremendous true reality, Freud is the father of the discovery of the unconscious individual , and Carl Gustav Jung is the father of the discovery of the collective unconscious , and Adler is the father of the discovery of the unconscious too and of the inferiority complex and guilty complex which are two great disease worst then cancer that affect 99 per cent of humanity , Freud psychoanalysis is focus basically on analysis of dream to determinate your day conscious life because when you go sleep and start to dream your one tent conscious were you pretend to deceive the society by acting a certain role is dispose , and during the dream you are totally unconscious so to see your real face you have to watch analyses the dream were the actor of one tent conscious is dispose, very acute intelligence Freud because when the actor of your one tent conscious that you show in your day to day life is dispose you cannot deceive , and your true face is show into the sleep dream , Gurdijeff a great mystic master used to make is follower drunk to give them drink and then watch at them, their behavior because when you are drunk you are totally unconscious and your one tent conscious that you show in your day to day to deceive people the actor is totally dispose when you are drunk and you can see the true face of the actor it work function fantastic, because me too I have experience to see the face of people when they are drunk and the actor of one tent conscious is totally dispose and they cannot deceive and you can see the monster in all is splendor . Carl Gustav Jung discover a tremendous true reality into the collective unconscious he say yes it exist the individual unconscious that one patient built since he born through all is life a huge accumulation of conditioning repression trauma wounds of the all life , but not only it exist a collective unconscious too were is store all past tradition of humanity all conditioning of family society, and old out of date expire religions, mode of behavior in society , superstition , fancy tale of religions that make the human being retarded crippled , and it affect totally condition totally all the life of the patient since he born on wards , the individual unconscious plus the collective unconscious were the neurons of the patient vector of thought believe emotion in a chain reaction into the brain, are feed with all wrong information conditioning repression wounds of inferiority complex and guilty complex , trauma that happen in the course of life and today days television cell phone computer conditioning give create into the patient a perfect hypnosis, conditioning that influence guide all of your life without any possibility of escape , unconscious the patient live a life artificial false of projection of outer sources of past out of date tradition dead of family of society system, and in short he live the life of other people , this in print of neurons false and artificial all lies build sideways the ego the unconscious of the patient that he rely to live is life and the ego too belong to society system religions old out of date expire , in short you live the life of other people influence condition guide by other people outer source , and your own original true nature is never live is totally neglected your inner being original your true nature original totally avoid neglected , it remain forever a seed not sprout a potentiality but it never ever become actual, your true nature of your inner being , Adler for example is a great figure of intelligence he bring to light a tremendous reality of the disease called inferiority complex and guilty complex that 99 per cent of humanity is affected influence, and this disease are worst then cancer I confirm it in my experience of life, people at first they have a tremendous inferiority complex and guilty complex create by the old out of date religions to dominate you better , I never understand how people have this tremendous disease called inferiority complex is a side effect of the ego unconscious mind they feel inferior, probable is the unlived life , and as you touch the wounds full of puss they get crazy mad schizophrenic as if you have touch a button that make them totally psycho pathologic , is incredible , and it trigger jealousy envy, two terrible side effect of inferiority complex and guilty complex that make the interaction with people almost impossible but I would say impossible . In the last century this 3 giant of intelligence have make of the world a better place a more true place, and were widely accepted from all the world, the world record nice the reality of the individual unconscious and collective unconscious hundred per cent , that by the way are just a partial truth , because then it exist the cosmic unconscious which goes back words into past life and many other layer of cosmic unconscious animals layer nature greenery layer rocks layer were the unconscious is totally asleep , bacteria layer up to the super conscious, non being incorporeal dimension less dimension core and source of the mystery of the universe and of life and death and of all duality , but that has never being record nice by humanity I tell you by the way is an immense huge reality of the cosmic unconscious all past life are imply into it and if you go through the inner journey into the cosmic unconscious you can revive and see all of your past life, and guaranty all the data of

those past life are there full alive , that is why I say all the time that the data the intelligence the essence the fragrance survive always and always death is only a changing of close of house , but the data the essence if you look into past life remain there intact because they are incorporeal they cannot be destroy by any death they survive, so be careful what you do because it will remain impress in the factual memory of the universe cosmos , almost for eternity to come . Anyhow this true reality of the cosmic unconscious is never being record nice by humanity but the individual unconscious and the collective unconscious yes as being record nice , neither the less nobody mansion this absolute true reality wen it come the time to do so to mansion , maybe because all fear are scare of this inner reality of our organic unity , also in the mystery school this attitude to fear the unconscious to be scare of the unconscious persist many people thousands I have seen when it came to face the unconscious run away and you never seen them again , because the unconscious individual and collective is a junk yard of neurosis disease in a way you see the monster that is hide within you and it scare people fear this encounter and run away , but if you run away at that stage you will never ever get heal again of the disease called unconscious , when it will happen in life again to encounter the unconscious you will be always stark there and run away , so is important to face the reality of the unconscious and go through the healing process until the really end, otherwise you will remain crippled disease your all life , I have gone through this process for 20 years of cleaning face the unconscious through hundreds of method and therapy, in the end it pay tremendous the healing and cleaning of the unconscious , wen instead of the unconscious within you is just an immense pillar of light consciousness , without boundary , but is heavy work one of the greatest method that I use and it pay tremendous in terms of healing and bringing light into the unconscious is Karma yoga, reunion through action it means , a Zen treaty but you need a community of Zen to use this process and is a very long process you react into the social life all of your unconscious conditioning wounds trauma, and the social Zen social situation burn them off one after the other , I was lucky because I have a situation of a Zen community were hundreds of great Zen master were present , but anyhow is important to complete the process of healing cleaning the unconscious up to the basement of it, and not get scare fear and run away if you run away you will remain crippled and you will have to begin always from the point that scare you and you run away .Anyhow something very important to understand of the unconscious is this that if you live a life of unconscious the consequence can be disaster dramatic , living a life of conditioning repression lies false information wrong notion, artificial a life of outer sources not yours a life of anthropomorphism a great disease ,were you are rootless or better rooted in falsity lies of personal God you will develop many serious disease like Alzheimer, the unconscious is the cause of Alzheimer because you rely all of your life in lies falsity conditioning repression wounds trauma, wrong notion wrong information fancy tale of religion, and your foundation of your house are bogus and wen a house as bogus foundation it fall apart sooner or later and that what Alzheimer is , loss of memory dislocation loss of capability of brain function , dementia senile , and you are like a tree without roots drying up to the sun , be very careful of the unconscious , and the unconscious it accelerate the deterioration of the cell of learning which into old people is a natural phenomenon that happen always , when you cannot learn anymore you stop evolving is like your DNA dry up and die before of is time , and the unconscious is the cause of this early deterioration of the cell of learning is an early death . True that men never learn anything never anyway it remain always identify with the ego unconscious mind and it only grows old and die , and he never grow up to grow up you need to be centered into your inner being and witness consciousness, and let go into consciousness awareness into non being body into the law of eternity, and be one intrinsic to it to eternity itself core and source of the mystery of the universe and of life and death and of all duality, were all forms duality time and space totally annihilate, is an incorporeal above beyond transcendental dimension less dimension canvas of the universe itself, were resurrection and immortality reside happen, always incorporeal is a conscious alchemy the resurrection not physical not material not gross . And last this humanity is totally identify hypnotize with misery suffering addict to misery suffering, fancy tale of religion it live in misery and suffering in anguish, due the influence of old out of date expire religion, it live in guilty complex that religions create to dominate you better it live in inferiority complex and this humanity is crippled , retarded and totally hypnotize,, that even if you go to them and tell them that life is a miracle of celebration a joy a playfulness, a mystery to love life is enlightenment life is eternal, death do not exist is the greatest fiction of humanity and weak up from the unconscious sleep, they will not believe the addiction to misery is so strong that they will go on living in misery anguish with this addiction hypnotize, such powerful is the unconscious behavior . For example they are society that are totally sadomasochist 50 per cent and 50 per cent masochist they like torture and be torture, they like tragedy they are addict to tragedy, that no matter that you tell them that comedy are more beautiful than tragedy, they are addict to misery tragedy , they like that drama of tragedy absolutely loveless society , I want finish like this the trade of deadly weapons of any kind as to be stopped abolish band as a crime against humanity this first principle of civilization as to be fulfill all over the world then civilization will flower never ever before, that the first principle of common sense of abolish band of all weapons as a crime against humanity as being fulfill , then we will have a conscious aware humanity and common sense will be a normal outcome consequence ......Angelo Aulisa

The new book Mysticism & Physics Chapter 11   Angelo Aulisa

Unconscious & conscious part 3

Hi friends , few important aspects of this important subject of the unconscious have to be added to give a better more complete picture of the unconscious & conscious, it will never being the all complete picture because I could go on for the all book talking of this subject but at least a few things more have to be say , so one organic unity built up an unconscious during his life , since he born until the present day that you are living your life, is a huge accumulation of conditioning repression wounds and trauma of situation lived, wrong notion false information of society system family, a bunch of lies, fancy tale past dead tradition out of date expire of expire religions , make up your individual and collective unconscious all of this unconscious staff are outer sources artificial projection into you they don't belong to you is the objective system society projected into you is not yours , and that is why that any criminal any monster psychopathic is a victim of the society of the system, byproduct of society system religions is the responsible of the society system expire religions that he belong , the criminal the monster psychopathic the sick man is the creation byproduct of society system religions out of date he belong , the all society system is responsible , the sick man demented , criminal monster psychopathic is just a victim of the society he belong byproduct of that society hence not responsible and you cannot condemn an unconscious human being is just a victim byproduct of that society he belong he has never had the alternative to become conscious they are no academy of meditation in the world that help the persons to become conscious hence you cannot condemn the unconscious persons, and please do not call individual because individual means an inner being that is all one indivisible with the core and source of the mystery of the universe and of life and death and of all duality , that is not the case of ordinary persons who live in sick society , in fact the all society system religions out of date expire family of that society is responsible for the crime that happen the tragedy that happen because the victim is just a byproduct creation of that society , you can condemn a conscious man but if is conscious no crime happen consciousness will prevent it , but you cannot condemn an unconscious person , person is a Greek word term which means mask he wear a mask of society unconscious , and he has never had any chance opportunity alternative to become conscious no academy of meditation exist that make the person become conscious , basic the all society system sick insane , is responsible of the act of crime the criminal is just a victim , if you want condemn you have to condemn the all society system is sick insane the all religions out of date expire are sick and insane, a bunch of neurosis lie superstition lies , fancy tale for retarded children that do not help the person in any way to become conscious, instead they make the person even more sick retarded crippled with guilty complex of being a sinners very vicious game they play the priest they make you guilty complex, inferiority complex, so they can dominate you and call you sinners , they divide reality into material mundane and spiritual they create a split into reality, so they can stand on a throne and call them self-spiritual and call you sinners, and they create guilty complex on you they should be banned all over the world the priest because the reality is that is all one, there is no division of spiritual and material mundane, they are two aspect of the same coins if you ask me I will tell you that the sacred is mundane, and the mundane is sacred, because life is sacred divine a gift of existence eternity itself to love and celebrate and rejoice , and the renunciation of spirituality the celibacy is a big crime against humanity, and the divine itself, that create monsters of priest pedophiles, and basically the renunciation attitude make a human being sick , insane , first of all the renunciation of the divine gift of life is sadomasochist attitude, and second your renunciation of a divine sacred gift of life from the divine existence is the renunciation of life itself , and very funny where do they go this people, priest who renounce the life they still live among us they weak up in the morning in this world and they need food shelter and everything , even if you run away in the mountain you still weak up among us and you need food shelter and all so is a laughing stock a sadomasochist attitude , and the priest they stand higher then you on a throne and call you sinners and make you guilty complex when in reality they are unconscious sick insane in sort psychological sick , please to the priest resign please resign enough is enough you damage humanity you make people sick insane, the within and without are one unique reality indivisible the without is an extension of the within and vice versa the within is an extension of without one unique reality, two aspect of a single coins, of the same consciousness awareness non being incorporeal body core and source of the mystery of the universe , and you the priest are not at all better than ordinary people you are the worst lies preachers . The people who rejoice life celebrate life are sane , and another aspect that I want underline is the inferiority complex , that unconscious built up , there are no superior no inferior being, all are sacred divine holy , the reality is uniqueness everybody is unique in itself original the rose is a rose the lotus is a lotus the marry gold is a marry gold, and there is no desire for the rose to become a lotus and no desire for the lotus to become a rose and no desire for the marry gold to become a lotus, they enjoy there majestic beauty in itself the rose is majestic divine early morning with dew drops in there petals shining to the early sun like diamond and full of rainbow color and fragrance , so is the lotus sprouting every morning to the first rays of the sun , so is the marry gold enjoying is colorful day light , and they have never thing of becoming one another , they are all original unique them self never ever a thought as cross there living being of becoming something else , the reality is uniqueness also for human being everyone is unique original in itself , different from one another different quality essence fragrance, and that make life beautiful the variety diversity of quality fragrance essence the biodiversity , there is no superior or inferior that are ego trip power trip of ignorant retarded people sick insane people , and finally beyond the forms there is a oneness of consciousness we are all one oceanic empty consciousness extension of one another, a sacred divine holy organism of oneness and every living being is sacred divine holy this out compassion for all of you for humanity love all of you ..Angelo Aulisa...

The new book Mysticism & Physics Chapter 12 Angelo Aulisa

Congratulation to noble price

hi Friends, I would like to congratulate and say thank you very much to the noble price commission , THANK YOU , the noble price for peace as being given to the right winner , I can , for rising consciousness about the extreme dangerous of nuclear weapons great , because the nuclear weapons as well hydrogen bomb atomic bomb and chemical weapons are the greatest dangerous of our beautiful green planet, that if delivered this weapons they can vaporize and destroy our planet in hours , and live radioactivity radiation that will go on for hundreds of years , is ashamed unbelievable that the nations of this planet earth have nuclear weapons to destroy this planet 20000 time the apex of folly and horror , to not count atomic and hydrogen and conventional weapons if we added it the nations of this planet have weapons to destroy vaporize this planet 50000 time is this not a folly a madness a big neurosis of humanity , totally crazy , the responsible goes to the old little men that is identify with old tradition of wars military violence , the old little men is very serious conservative of old out of date expire tradition of wars , military and building deadly weapons and make wars massacre killing people doing bloodshed destroy the planet , in the past ancient primitive barbarous unconscious humanity they have fight ten thousand wars and kill millions and millions of people destruction and killing and bloodshed was, the way they interact between nation , they have no better way or better things to do then fighting wars and killing and destruction , they were primitive barbarous unconscious infantile childish ignorant , but fortunate their weapons were ancient and primitive so they create less damage local damage , but they were insane neurotic mad anyhow , not evolve , the DNA of human being was in a phase of evolution not evolve yet , partially we can understand the primitive unconscious men was ignorant insane neurotic ignorant not evolve , but coming to our contemporary age is incredible to see that the old little serious men of our age , seriousness is a disease of the ego unconscious mind only a serious men can fight a war ,he still look at the past ancient out date tradition of wars to interact in the present with other nation , today days still prevail the law of the jungle that the nation who as the most deadly weapons is the most powerful and the interaction equilibrium between nations is still determinate by the weapons by the military power , this is unbelievable this humanity seems that as not evolve in intelligence consciousness awareness at all as remain barbarous primitive unconscious, the DNA of the little old men as remain unconscious barbarous , but today days the dangerous is that they have weapons incredible dangerous like nuclear hydrogen atomic that they can destroy our green planet in hours vaporize putting an end to our growing so called civilization forever , is like to give a gun in the hand of a child that he don't know for what is use for , And this world humanity today days is absolutely unconscious asleep , either the individual is unconscious asleep either the nation live through the collective unconscious which is nothing else that all the past tradition of humanity store into the collective unconscious ,system family out date expire religions included in it , in fact today days we have a flowering of dictators tyrannies in many nations of the world and this is even more dangerous because to give in the hands of tyrannies dictators this deadly dangerous weapons like nuclear atomic hydrogen is dangerous like hell, sooner or later one of this dictator tyrannies which they have all of them an abnormal ego unconscious he will push the button putting an end to our beautiful green planet , the flowering of this dictator tyrannies regime they are many in many nations , in Africa , in middle east , in Asia , in Russia , in Syria , in Turkey , in North Korea ,in south America , in Unite State even , and many country in the world that show a mask of democracy is just a mask beyond the mask there is regime dictator and tyrannies , that don't allowed freedom of speech or freedom of choice and freedom of living the life in freedom basic real regime dictatorship tyrannies , is a big dangerous for the entire world , this regime dictator tyrannies flower because the people the society of the nation live unconscious asleep, into the collective unconscious of past dead tradition which they have to be abolish completely , when a person don't know who he is don't know where he come from the don't were he is going don't know about life and death don't know about the organism of the universe that he live in , he is scare and fear the immense universe he live in, is easy to exploit and to make a slave of him , the little men of ordinary society is conscious retarded or say spiritually retarded due the conditioning of expire out of date religions that preach superstition fancy tale lies from the begin to the end , a bunch of neurosis and lies a night mare of the ancient past humanity , and so the small old little men is anthropomorphic retarded , unconscious asleep, scare fear full ignorant , and he need a father figure first of all in the sky the fiction of God but not only he need a father figure also as a leader of nation a dictator a tyrannies a regime, fascist Nazis whatever as far he feel safe , this is the failure of intelligence of human being this is what we are witness this day all over the world , the flowering of regime dictator and tyrannies , when the reality I repeat always that a leader can stay in power for five years of legislature no more even if is good and intelligent, after five years of legislature he as to resign , because in five years he build an unconscious habit to power and he became a dangerous of all population, because absolute power corrupt absolutely , because today days there is not a single leader conscious they are all of them absolutely unconscious and dangerous like hell , five years should be a limit for any legislature in the world to avoid the flowering of tyrannies and dictators , is incredible that the world all the way suddenly as plunk hundreds of years back words into regime in dictators tyrannies, starting from unite state were I can see clear that people cannot have freedom of speech and freedom of choice otherwise they are fired by the leader of the country incredible , but the primary reason as I tell you before his because people are lost unconscious, spiritually retarded scare and fear full and live in ignorance, so they need a father figure as leader that give them kind of safety but this is ashamed , is the failure of human intelligence , that is why I say always wake up from unconscious asleep , get centered into your inner being and witness consciousness in your true nature get roots into your inner being and into your inner consciousness awareness, into non being incorporeal body into the law of eternity core and source of the mystery of the universe and of life and death and of all duality , because once you are rooted into your inner being and simultaneous into all existence nobody can enslave you anymore with fancy tale or superstition or out of date retarded ignorant speech like the tyrant are doing , once you are rooted into your true nature of your inner being you are rooted in love because love is the intrinsic nature of your inner being, you in essence are love intelligence compassion peace, a miracle of celebration , love is not an outer phenomenon is an inner phenomenon in essence into your inner being you are love itself, and once you are rooted in love you can begin to share your love with the all world and the all universe , and nobody can enslave you anymore , once you are love in essence and you share your love you laugh at the tyrannies dictator , weak up from unconscious sleep once u are rooted centered into your true nature of your inner being simultaneous you are centered into the all existence, into your inner universal consciousness and awareness into non being body incorporeal because time space forms duality totally annihilate into the law of eternity is just a great relaxation a great I am ness the law of eternity, and the core and source of the mystery of the universe and of life and death and of duality, the canvas beyond above transcendental bigger and vaster then the universe itself, the core and source were the universe is display paint . Relax and go for the inner journey the path meditation , silence ,love painting singing dancing laughing whatever take you beyond the mind into no mind is the path , and once you know who you are were you come from where you are going, melted annihilate into the universal body consciousness awareness, into non being body into the law of eternity nobody can enslave you anymore , you know the real freedom into the law of eternity the real relaxation , and basic you know your inner being you are rooted into truth you are enlightened free forever, immortal the resurrection as happen forever . Reclaim your freedom you have nothing to lose but the chain of your conditioning repression , Basic I am for the abolish ban stop of all weapons nuclear and conventional all of them is a crime against humanity to build them, to have them, and the band stop abolish of wars and military as a crime against humanity, is a crime to hurt kill an extension of your consciousness every living being is sacred holy divine , and this will be the first step towards civilization, towards fulfilling the first principle of civilization which has never exist up to today , only talking dreaming , this earth the lotus paradise this body the consciousness awareness , that is it , and thank you to the noble price commission also for the price on gravitational wave this may trigger to shift the attention interest in the exploration of the universe the real direction orientation of humanity and give more resources to physic and science , we need a strong physic and science , the world of the future for 90 [er cent will be guide objective by science and physic , 10 per cent by the inner science of mysticism the inner reality of men by meditation consciousness awareness , we need urgent a new dawn of civilization intelligence empty consciousness label less content less adjectives less of awareness , no partial interpretation of the little old men consciousness is neutral to any interpretation and neutral to gender color race age , is for everyone to rediscover it and be one with it is just a fundamental law of the universe, the intrinsic fabric of life just a quality a pulsation a creativity intrinsic to the universe physic called it Boson x the quality that keep hold together the all universe and confer Mass to particles atoms matter without the boson the all universe disintegrate instant that is what the mystic call consciousness since thousands of years two name for the same quality fundamental law of the universe .....Angelo Aulisa

Published on October 09, 2017 00:49 • 21 views

The new book Mysticism & Physics Chapter 13 Angelo Aulisa

Congratulation to Nobel peace prize part two

Hi friends , so congratulation again and again to Nobel commission price , and congratulation to I Can , the organization who win the peace price, your work play is sacred divine holy for this planet earth , the dangerous of nuclear atomic hydrogen chemical weapons is an horror a tragedy for the world to afford, is very important to rise consciousness on this subject of deadly weapons what they can do to the planet earth , this deadly device weapons can vaporize destroy our green planet in hours and live behind radioactivity radiation for hundreds of years and put an end to our growing so called civilization forever , because people are not conscious of the dangerous of this nuclear weapons, they show it as showing piece with proud are totally folly mad neurotic crazy hypnotize unconscious asleep , this absolutely deadly weapons are not toy but are the end of planet earth as a green planet and the end of humanity on earth, weak up from unconscious asleep right now , and the contribution of I CAN is immense and right as being the Nobel peace prize for them I am very happy that humanity as award you I CAN thank you for your work . I am for the abolish ban stop of all nuclear atomic hydrogen chemical and conventional weapons, deactivation destroy the existing one and ban abolish factory industry of weapons that are build new one, they are doing a crime against humanity and a big crime against consciousness awareness intelligence of human being , and I am for the abolish ban stop of military and wars all over the world when there is no weapons as they are ban abolish also the military they will be useless not needed they can dissolve into a peace force without weapons all over the world and the world, to recover from calamity of all kind, starting from climate change , and no military the consequence will be no wars finish of this ugly out date expire tradition of wars, it belong to the past ancient humanity barbarous unconscious primitive out date expire childish infantile neurotic mad crazy , the unconscious primitive ancient men as fight ten thousand wars in 5000 years , and kill millions and millions of people, destroy entire nation and population with bloodshed tragedy horrific , they were interacting through nation with wars destruction and killing, this attitude of past tradition out of date as to be abolish ban stop right now, and the new way of interacting between nations will be intelligence peace consciousness, awareness compassion love harmony, balance , we measure our intelligence not our weapons the new men as evolve incredible in the last thousands of years, the individual and collective DNA of humanity demand urgent this convergence of evolution mutation, and quantum leap towards civilization , the first principle of civilization to fulfill is a world without weapons, without military without wars, without violence basic a conscious aware world ,were the individual and collective unconscious source of crime violence and conflict as being annihilate totally, and only a pillar of light empty consciousness awareness is left, either in the individual unconscious or collective unconscious this transformation of bringing light into the unconscious goes through the abolish and ban stop of all tradition of the past included all out of date expire religions , the ban and abolish of all kind of weapons as they are a crime against humanity and civilization the ban and abolish of all military and wars all over the world , We will create for the next hundred years a stronger police more police station everywhere , and only police will be allowed to have light weapons to protect and monitored the transformation of the planet towards consciousness awareness, towards a mystical union with the universal body, and non being body incorporeal body were time space forms and duality of mind totally annihilate, that is why incorporeal or call it the law of eternity core and source of the mystery of the universe and of life and death and of all duality, the ultimate canvas were the universe is display paint vaster bigger beyond above transcendental to the universe itself, a great mystical union of humanity with the core and source of the universe , a great organism oneness of humanity with the law of eternity , Men in essence is eternity itself immortal resurrect, men is a oneness mystical union with eternity, and wen this alchemy as happen the unconscious sleep is gone forever and men is enlightened free, he live the real freedom in essence into eternity into an open infinite relativity, without boundary . And thanks to the Nobel price commission for the price on gravitational wave this may trigger a real interest in the exploration of the universe and give more resource to physic and science ,very important because they will be the leader of the objective world they will change the face of the planet earth and they will change the climate change only physic and science can face this task if is not too late to change the climate change, and they need immense resources , This are the orientation direction of the world for a new dawn of civilization, intelligence empty consciousness, awareness, into non being into the law of eternity core and source of the mystery of the universe , And this mutation transformation will change the society system of the world, first of all we need academy of meditation all over the world, were the persons can have the alternative chance possibility of becoming conscious aware , then we will have society system of individual, and not society of persons who were an unconscious, mask of society person means mask in Greek language , and individual means an organic unity who live in mystical union indivisible with the core and source of eternity non being , and when we will have society of individual a transformation from democracy out of date expire is just a name for exploitation for corruption democracy is just a nick name to cheat people , democracy as exist at the time of Plato in Greece for 30 years in Athens that people meet in center of square and discuss talk direction orientation of Athens animate , then it finish then and there , what we see today is the totally failure of honesty democracy is over since 3 thousand years , the new way will be Meritocracy one by the merit of is action in life earn the possibility of being a leader and he will have to show certificate of academy of meditation that prove the man unconscious free , that is the only guaranty that he will not exploit cheat people later with pent up of unconscious behavior , the unconscious is the source of crime and corruption and exploitation , society system of individual centered into is inner being and witness consciousness awareness , and society system guide by Meritocracy , and every leader can be in power for a legislature of five years no more , because everyone is unique into itself and as different quality essence intelligence that can bring innovation in different way, because we are seven billions of individual in this world and everyone intelligent and with different essence quality that can bring new innovation as the other could not do , and five years is a limit also to avoid tyrannies dictators that build up unconscious behavior to power , power corrupt and absolute power corrupt absolutely , ,This the new guidelines for a better world that you will find better into a book called the world constitution write by an author called Angelo Aulisa , and more the abolish and shut down stop of prisons as a crime against humanity , they will be substitute by academy of rebalance and harmonization of the unconscious of the persons , with therapist groups and meditation method of all kind , and in the evening the persons they can came back home to his beloved one , the prisons were primitive barbarous method of punish the persons unconscious that by the way you cannot condemn an unconscious persons is not responsible he as not have the chance alternative to become conscious , you can give this alternative through academy of re balancing re harmonization of is unconscious, into realize is inner being and witness consciousness and then no crime can happen anymore consciousness will prevent any criminal action , anybody within is sacred holy divine one in mystical union with the core and source of life is only a matter to re balance re harmonization, not of punish anyone , and last we will have a morality relative to the moment and circumstance, morality is the shadow of your inner being once you are centered into your inner being morality will be just a relativity of your inner being a shadow of your inner being, sacred holy divine in every situation circumstance they will be a natural spontaneous response of your inner being, different from one situation to another morality is relative to the moment of the situation that unfold and just a shadow of your inner being, morality is not relate to any credo religions expire and out of date and abolish in a new world constitution no more lies no more fancy tale no more anthropomorphism a dangerous disease that has brought the world into a constant conflict and wars and destruction and killing abolish ban stop, not allowed to spread lies hypocrisy hypothetical fiction of God , not allowed humanity as come of age we know our core and source of life for real the truth of life and death for real the truth beyond the duality of mind unconscious , we know who we are where we come from where we are going and most of all we know the fiction of death that do not exist, what real exist is eternal life in essence a human being is eternity itself immortal resurrect, awake enlightened forever and ever so this the guidelines for a better world of peace celebration of love, compassion of intelligence evolve of civilization for real without weapons wars and military, of meditation of consciousness awareness , for a world basic that live conscious aware in mystical union with the organism of the universal body and the law of eternity itself ....Angelo Aulisa

Published on October 10, 2017 01:49 • 21 views

The new book Mysticism & Physics Chapter 14 Angelo Aulisa

Woman & enlightenment

Hi friends , I was thinking to write a different book and I will do , my subject is physic mysticism , science inner and outer , and I was on the way of doing it , but the objectivity spark everyday new debate through various media , and I see that humanity is stumbling in darkness ignorance , and out of compassion and basically to be contemporary I have to sheer light on various subject , this subject of woman and enlightenment is very important extreme important , I have write chapters through my books about this subject were already say much about it , but here it come a new opportunity to added something very important about it , Woman through the century through thousands of years have being mistreated very badly , have being repressed very badly , have being deprived of their own basic right all over the world and have being treated in the most ugly way basic there is no words to describe the night mare that woman have going through down the century , woman have being shakedown beet , have going through the most horrible violence men the maschilist chauvinist, men as behave terrible against with the woman absolutely inhuman , The credit and responsible of this mistreatment of woman goes in first place to all out date expire superstition false ugly religions all over the world , all religions they dont consider woman as real human being, many religions all of them I would say think that woman are soul less that they don't have a soul ridicules and absurd, sorry they are all of them the religions demented monster psychopathic , a night mare of the past ancient humanity a bunch of neurosis lies , this behave of the religions against woman is due an inferiority complex that men and religions has against woman , they fear their immense intelligence and power , so they don't consider woman human being with a soul absurd unbelievable incredible , an injustice endless a crime against humanity, perpetuate by the all out of date superstition religions of the world , many time I ask myself how humanity allowed this religions to be still there is sheer hypocrisy , woman are half population of the world the salt of the world the poetry of the world the most intelligent sensitive intuitive of the world, the beauty of the world and they are sacred holy divine as any other living being , and consciousness awareness is neutral to gender color race age , any woman has an inner being a soul any woman can be one intrinsic to consciousness awareness as any other human being any woman can be enlightened as any other human being, and basically in my experience of life I have witness more woman enlightened then men , they have more the basic quality essence sensitiveness intuitiveness quality to be enlightened then men and basic they have at least ten time more power energy than men to transform into consciousness awareness creativity enlightenment , a woman is multiple orgasmic a man only one , a woman is a womb receiver absorb existence consciousness a men is a rock like worthless , he repeal consciousness by is rocky attitude aggressive violent not sensitive not intuitive also due is diet meat eater cannibal , and so a woman as an inner being a soul a woman is ultra-consciousness awareness, a woman is easy enlightened . Now all religions around the world are doing a crime against humanity creating guilty complex and inferiority complex in woman of all kind ,because they are demented senile insane , I want tell to all the woman around the world to reclaim your freedom your individuality, your soul inner being ,your consciousness awareness and the right to be enlightened because this is the truth of consciousness that is neutral to gender color race age , consciousness is a basic fundamental law intrinsic to the universe the very fabric of life of the universe just a quality a pulsation of love intelligence , a creativity intrinsic to the universe expand everywhere and nowhere in particular into the universe, is a tread that run through within all living being and forms of the universe and make them alive animate them, physic called consciousness Boson x the quality that keep hold together the all universe and confer Mass to particles atoms matter, without the boson the universe and forms of it disintegrate instant , mysticism called the Boson consciousness since thousands of years, two name of the same quality, neutral anyone can be one intrinsic to it through various path love for woman is easy to be one with consciousness, through love or silence meditation , painting , singing , dancing , laughing , sculpting , any path that bring you beyond the mind into no mind into your inner being and witness consciousness is the right path , avoid denied abolish the out of date expire religions that create in woman only guilty complex and inferiority complex are you nuts to keep them on they should resign right now demented senile insane expire religions sorry but enough is enough, of anthropomorphism a terrible disease that as keep the world divide in conflict wars bloodshed destruction since thousands of years , like a mirror broken in thousands of pieces and every pieces claim that is the truth bullshit sorry they are all of them a bunch of neurosis lies superstition insane senile demented a night mare of the ancient past unconscious primitive humanity, the truth is the empty mirror all one, the truth is the empty consciousness no labels no adjectives no contents , they were all partial interpretation of the little old ugly men . So basic anyone can be in essence consciousness awareness, one intrinsic to it woman at first place. I want that you reclaim your dignity and human right totally and right now, to all the woman of the world , the first step is to abolish ban all religions of the world that they spread lies and lies not only about woman but also and most of all about hypothetical fiction of God is an hypothesis at the most an hypothesis, open your eyes is a disease called anthropomorphism a terrible disease that create conflict and wars since 5000 thousand years , all the religions they should they must resign, if we want a better world that reclaim all the intelligence evolve in thousands of years and a true authentic world real that live factual in reality, that physic science mysticism have shown to us every day every split second based on experience and experimentation , Not like the expire religions based on imagination and hypothesis, no one has come back and say that as seen God it do not exist is an hypothesis, all the religions are archaic with traditions of the past stark with them and they are so childish infantile fancy tale so absurd and senile, and they pretend dictatorial that we should believe in them , no dear senile priest enough is enough you are destroying the world creating conflict and wars with your absurd senile archaic tradition, we abolish all of them and we ask please as a favor to humanity to all religions to resign right now , physic and science have demystify all of them , is important that with the resignation of all religions that have being an hypnosis for humanity for century we should come in propose a new approach path way for humanity , science and physic will lead the objective reality of the planet for 90 per cent, and ten per cent mysticism the inner science of the mystery of the inner reality of men , meditation inner being witness consciousness , inner consciousness non being incorporeal, because time space forms duality of mind unconscious totally annihilate into non being, into the law of eternity begin less endless never born or die, that the meaning of eternity, a new dawn of civilization intelligence meditation empty consciousness awareness , into non being incorporeal dimension less dimension called the law of eternity core and source of the mystery of the universe and of life and death and of all duality, an infinite canvas bigger and vaster then the universe itself were the universe is display paint . Is important to give to humanity with the resignation of all expire out of date religions that were a deep hypnosis opium hypothesis lies from the begin to the end senile insane , an alternative a new way a new path true and authentic a new era of civilization intelligence, meditation empty consciousness awareness, love peace compassion balance, harmony playfulness celebration , bliss , subtle ecstasy , joy , so the abolish resignation ban of all religions out of date expire hypothetical superstition false pseudo senile insane , and the introduction of a new dawn of civilization intelligence, consciousness awareness urgent needed now , this is the introduction to woman and enlightenment is gone be a long chapter this is part one .....Angelo Aulisa....

The new book Mysticism & Physics Chapter 15 Angelo Aulisa

Woman & enlightenment part two

Hi friends , so the biggest hindrance for woman liberation are the out of date expire religions that create in woman guilty complex and inferiority complex , and as I say before they have to resign and be abolish ban forever they have done their job somehow they have keep humanity keep on in an illusion of poetry false and lies and superstition of course fiction of God of course but they have done this work unconscious and humanity somehow as survive, now is time with kindness gentle out of love for humanity to step down resign for all of them and let humanity evolve in intelligence and truth authentic real , I am not hungry at them or at anyone I even say thank you but please get out of the way enough is enough . And so woman need compensation for the ugly insane behavior that humanity and out of date religions have had against them big compensation , now we can understand partially the mistreatment that in the past ancient primitive barbarous humanity woman have gone through, they were totally unconscious primitive society , but we cannot tolerate this ugly behavior of religions expire out of date any longer , in our contemporary age of 2017 , woman have a soul an inner being and they are conscious aware and they can be totally enlightened because consciousness awareness is neutral to gender color race age , and more the out of date expire religions have and are doing a big crime against humanity and woman and men through their approach of renunciation to the sexual life to the sexual energy with celibacy , Celibacy is a crime against humanity a repression of sexual energy that create all sort of monsters, psychopathic priest, pedophilias , the attitude of religions that sex is something to repress dark taboo dirty , is ignorance bigot insane a big lie an ashamed ignorance ,try to conceive so we are born out of a dark dirty thing not out of love I ask , that create repression conditioning so ugly and insane senile people , Celibacy is a crime the sexual energy of a human being is natural spontaneous clean and beautiful phenomenon, celebration is the way the path that create love life , the sexual energy of a human being is like a river flowing to the ocean you cannot build dam of repression conditioning without create perversion and perverted people , to interfere in the sexual life of a human being is a big crime and religions are perpetuating this crime since ever with absurd explanation nobody is impotent, everybody is potent if you repressed condition such a huge immense river of sexual energy it will take side ways of perversion , of all kind I don't mansion the catalog of perversion , instead I ensure you that if you let the river of sexual energy flow spontaneous natural free of conditioning and repression to the ocean, no perversion will take on , human being will love live through the natural way of opposite complementary , of thesis woman antithesis men synthesis of love the dialectic law of nature the dialectic law you can call also the law of nature of opposite complementary , Is a well-known fact that dogs and cats and other animals become guys when they live in captivity when they live in domestic situation that bring repression and conditioning to them, otherwise an animal's cats docs etc... are never ever guys , I say this to mean that is very important to not interfere in anyway in the freedom naturalness spontaneity of the sexual energy of a human being, first of all is a crime second if we want a better world sane and healthy man and woman , expression is the magic word natural spontaneous in freedom is the magic word , But the religions with all their archaic tradition stench out of date, interfere and repress conditioning every aspect of life and sex is the primary thing that they interfere and they create a mess a chaos of perversion into humanity, I repeat celibacy is a crime, and celebration the expression the path the way, for a better world sex is a natural spontaneous beautiful phenomenon to enjoy and celebrate to rejoice . The interference of religion into the sexual life of society and system as create a very perverted humanity , the way that society are actually is insane senile repressive a full conditioning, without any out let without any window that what is the family the most repressive institution of society , the end of freedom no out let no window an absolute repressive way of living the life , I call the contract of marriage legalize prostitution sorry this is the truth , unless is a love marriage , but even then the way society lives soon it will become a cage a prison cell , I see a society system of individual centered into their own inner being and witness consciousness that they don't fear loneliness, but are alone individual that means all one with the organism of the universal body and consciousness and awareness, into non being into the law of eternity , and to be alone is sacred divine a subtle ecstasy, and when you are all one alone you are one with eternity itself the core and source of the mystery of the universe, were all the enlightened one that have had a resurrection reside so you are with Jesus Buddha Mahavira Basho Lao-tzu OSHO Body Dharma and thousands of enlightened one, when you are conscious you are focused into the objectivity when you are aware you are relation less unfocused relaxed into the law of eternity begin less endless never born or die in great company, you are together with the all existence is a great bliss subtle ecstasy divine , instead people fear loneliness and they get married because of this tremendous fear of loneliness , humanity fear is scare of loneliness and death like hell like nothing else , instead in a society of free individual the marriage and the family will disappear because we will have a society of healthy individual, and consequence an healthy society that live in freedom and love will be a sharing out of abundance , basic one into is inner being in essence is love and love is an inner phenomenon not an outer phenomenon, and one is centered into is inner being and witness consciousness is love itself, and he can begin to share is love with the all universe and planet , the family and marriage will disappear totally the family is source of conditioning repression is the end of freedom is enclosing yourself in a small unity of people no window no out let of freedom of any kind , instead the individual lives into community Buddha field academy mystery school Zen field of thousands of awakened one men and woman of course from all over the world he live into a micro Cosmos of sharing of experiences love feeling dance of meditation, and basic celebration expression of all kind event of all kind he will have an abundant life of celebration and sharing experience , he will have an universal mind , open to all experience , rich , basically I see the future of humanity into millions of Zen field academy were the children will have different mother and different father and all uncle , because the old out date family is to possessive for the children a real father will repress the child with attachment and possessiveness he will crash is freedom , instead a Buddha father will enhance all the quality of the child he will live with the child as a friend and when a child as many father he will have a universal mind free of conditioning of possessiveness and attachment that crippled the child, you see finally we belong to the universe to consciousness awareness to eternity, the claiming of the real father and mother that is my, is possessiveness attachment that condition repress the child irreversible , A new era for real millions of Zen community field, academy all over the world meditation consciousness awareness the inner orientation direction of any Zen field academy, and basic any Zen field academy is a center of celebration life event music dance , were all the enlightened one share their experience life , the real dream for all the world then this world the lotus paradise this body the consciousness awareness , here end part two of woman and enlightenment part two is gone be a long chapter soon will follow ...Angelo Aulisa

Published on October 12, 2017 11:18 • 37 views

The new book Mysticism & Physics Chapter 16 Angelo Aulisa

Woman & enlightenment part 3

Hi friends , you maybe puzzle because I talk have a vision of a world totally new a real revolution transformation of all the field of the world , a world without weapons without military without wars , without organize religions , without marriage without family , and without nations , they will be functional nation like the post office functional govern of nations , all related to a super strong UN , with a single constitution for the all world , a global world , with a single passport the UN passport for every individual , I repeat they will be functional nations govern but all under the guidelines of the UN the unite nation will be the coalition of all governs each functional governs will have members into the united nations, but the constitution will be one world constitution that all functional governs will have to subscribe and sign , if they want seat into the unite nation a strong UN govern for the all world were all the nations will have equal power of decision , they will be no weapons no military no wars , no boundary of nations , no religions no society of family no discrimination of gender color race age , physic and science will lead the objectivity of the world for 90 per cent , and ten per cent will be led by a new dawn of civilization intelligence empty consciousness awareness, into non being incorporeal dimension less dimension, were time space forms duality of mind totally annihilate that is why incorporeal, into the law of eternity core and source of the mystery of the universe and of life and death and of all duality, ultimate canvas of the universe, beyond above transcendental, to the universe itself actually were the universe is display paint , sacred divine holy , a new humanity were for the first time civilization will take plays, were civilization will be a reality for the first time since ten thousands of years, the triumph of intelligence, the triumph of the DNA of human being the greatest convergence of evolution quantum leap of humanity, were all human right are sacred respected a new era , ha ha ha take it easy is just a dream I am not thinking that this will happen soon, the only think that I mind is that maybe I will not see this great new dawn of empty consciousness awareness, civilization intelligence meditation mean while I am in the body, but I promise I will enjoy it formless into the universal body into consciousness into awareness into the law of eternity this I promise , because this transformation quantum leap convergence of evolution of humanity will take hundreds of years to happen, and maybe yes and maybe not because a total destruction through nuclear atomic hydrogen weapons is always at the gate, a possibility , if the world will be not be destroy by the old little men, then maybe it will happen , but I repeat for me is just a dream I know that I am shouting in desert and nobody is hearing me, but I am an incurable dreamer and I like to spell the truth this for sure , I say this thing because suddenly I have think that my vision is little beat too much for this unconscious humanity ha ha, I know it very well but I am an not serious individual I am very playful , I say the thing that I envision but I don't pretend that people understand or comply with me , to me freedom is the first principle so I understand that you all friends have the freedom to laugh about the things I say absolute freedom , but remember I spell the truth all the time without hypocrisy, this you can be sure about because basically there is nobody within me that is saying anything, is just a divine flowing of spontaneity , it is say that Gautama the Buddha spoke for forty years without utter a single words , now how this can be , same I say to you all friends I have not spoken a single words is all a divine flowing in the spoor of the moment , and if what I say happen beautiful for you all not for me because for me as already happen everything, I am happen and that is more than enough, I am happening every split second and that is the miracle , someone ask Buddha you are a Buddha so please perform a miracle to show us your divine power , and Buddha answer I am the miracle . me to I answer I am the miracle , that is it and all , if whatever come into light in my writing happen you take it is all good for you , because I am no more since more than 35 years I think it was 20 February 1982 , how many years are I don't know that I got enlightened and disappear since then a new presence fresh every split second a consciousness awareness totally empty fresh every split second , so I don't mind if humanity consider too much my writing actually I don't care at all , but it as to be say . So let come back to the subject of woman and enlightenment , the society system of today days still goes on to repress conditioning woman in all the ways , for example what you call family or marriage is an history of domestic violence, either psychological or physical the woman mostly are repressed crash shakedown blackmail in the most ugly ways, the marriage is an history of enslaving one another, actually the woman have the worst part of being enslave by the husband but not only, also the husband is enslave by the wife , the best couple he his sadomasochist and she is masochist the failure of civilization , we the mystic call the marriage legal prostitution with the sanction of a contract , because that contract is the end of freedom love and spontaneity , sorry , That is why I propose a totally different society system were love substitute marriage , love will be the tread that band together all humanity and human being , one live together until love is present alive when love fade, wider away we change ,we are seven billions of individual you can have different story experience instead of torturing your wife , or vice versa , and your society system are insane sick with shakedown blackmail the woman in the most ugly ways , there is a huge population of sex slavery were woman have to sale there body to live and there is a huge exploitation of woman shakedown of woman that if they want work they have to go through the red light room of the owner, and do all kind of sexual story perversion , since ever, this of sex slavery and shakedown for working in any place of society are ancient tradition of the world, they are old as the world is old , nice that in America they are rising this debate trying to bring to light this wounds of collective unconscious , well you have to know that they are hundreds of millions of owner of any kind of economic power, which they require in regular base that the woman who want the work the job, first as to satisfy the owner the boss with sexual activity then she will get the job the work , is a tradition ancient as the world is ancient , very ugly of course but woman have being are exploited in thousands of way , is a crime of course , but that the way it is all over the world since ever , the collective unconscious of humanity is full of wounds horrible drama and trauma of such shakedown blackmail of woman, some horrible some dramatic some bloodshed but that the way it works and his . But I am surprise that the American are so sensitive to shakedown of woman on getting work job , that they have to be rape, in way is a crime ugly I agree , but the one who perpetuate this crimes are totally unconscious either the woman that is the victim, either the men that is the criminal booth are unconscious and victim of your sick insane society, sexually repressed by religions family system , and unconscious persons cannot be condemn , are just victim byproduct of your sick insane society , product of your society , but the American that are not sensitive to weapons and massacre , they should create debate on weapons abolish and ban totally, and shut down of factory of weapons abolish ban as a crime against humanity , because if you want persecuted this shakedown on woman for working and getting job you will have to persecuted at least one hundred millions American people, that are doing in regular base this blackmail shakedown of woman and one billion around the world, is an ancient tradition of the world that why I repeat always that all tradition are to be abolish ban, we live here flowing moment by moment in the present in consciousness awareness tradition are out of date expire dead, of past get reed of all kind of tradition forever ... ok this part 3 end here follow more next ... Angelo Aulisa....

The new book Mysticism & Physics Chapter 17 Angelo Aulisa

Woman & enlightenment part 4
Hi friends , so woman have being exploited are exploited in thousands of ways , until the society system around the world remain and are unconscious and live in and through the collective unconscious were all tradition of the past are store were all out of date expire superstition pseudo religions are store were all system of society are store there is no hope for woman to come out of this ugly hypocrisy perpetuate against them, also for men is the same in minor scale , until men lives into the individual and collective unconscious there is no hope of seen a better world because the collective and individual unconscious is the source of all the crime and violence , for example until humanity will lives into the collective unconscious and individual unconscious the shakedown blackmail of woman will go on, I would say it will flower more and more , and is very ugly situation for woman a crime endless , and until the society and system will have the family as center of their system society, the sex slavery trade will go on I would say will flower because the family is responsible for the side effect, of sex slavery is a very ugly situation and big crime , wen this humanity will mutate change transform itself and have a quantum leap a convergence of evolution into live conscious aware, and that means a quantum leap into the vertical dimension of deeper and higher of inner being and witness consciousness through meditation , and that means a quantum leap into love because the basic intrinsic nature essence of your inner being is love you in essence are love when you live through your true nature, then wen love is the inner tread that band us all , the shakedown blackmail of woman will drop finish then the sex slavery will drop finish because no more family no more sex slavery, is natural equation of living conscious aware in love, with a love tread for all and everything , until this humanity will live in the horizontal dimension on the surface superficial from abc, xvz , then the collective unconscious and the individual unconscious will be the normal trend with tragic consequence for woman and for all humanity general . The transformation the convergence of evolution quantum leap of humanity dear friends is basic love, to substitute the dark age of the unconscious with inner true nature of love which a human being in essence, is living into the dimension of being love is the inner intrinsic reality of human being , living into the dimension of matter and having the unconscious primitive barbarous violent and wild is the reality of the surface superficial objectivity , with love everything is transformed no family, no shakedown no sex slavery who want be a sex slave if love is at hand the reality nobody , love will substitute of the marriage contract, love will substitute the family love will substitute the shakedown love will substitute the sex slavery trade, it will be a better world in first place for woman and basic for everybody , a bliss to live a celebration a subtle ecstasy a divine freedom a joy a rejoice into the miracle of love and life, but this step quantum leap into the vertical dimension of deeper and higher of here and flowing moment by moment into the present into consciousness awareness is require absolutely . Now a few guidelines for woman and enlightenment, is ashamed that many out of date religions consider woman not even spiritual without a soul, incredible unbelievable they are all of them the religions demented senile insane they are keeping on tradition of the past that their religions have preach down the age lies absurdity superstition neurosis of ancient past tradition a night mare a bunch of neurosis , but I tell you many time that consciousness awareness is neutral to gender color race age , is just a fundamental law of the universe the intrinsic fabric of life just a pulsation of love intelligence, a quality a creativity an inner tread that run through all living being and forms of life, expand everywhere and nowhere in particular into the universe and any one can be one intrinsic to consciousness, woman at first place consciousness is neutral to any interpretation of the little old men, is label less adjectives less content less just a divine sacred pulsation of love intelligence vitality in fact consciousness is the source of love and intelligence , the intelligence that people call intelligence is just information you have read a few books seen a few documentary on television look into the computer day and night and you think that you are intelligent forget is just information, into your brain can be store all the library of the world and then you think that you know and you are intelligent, forget is just information intellect , intelligence is in the spoor of the moment in the split present fresh moment by moment, and the source is into your inner being and consequence into your inner consciousness were the essence data fragrance of thousands of enlightened one have annihilate, for an eternal resurrection in fact sometime I channel Jesus sometime I channel Buddha sometime I channel Mahavira sometime I channel Lao-tzu sometime I channel Basho sometime I channel Rinzai , and basically all together wen I am totally relaxed into awareness basically all together is an immense source of intelligence, light empty consciousness but if you are skill you can chose at your pleasure, the resurrection is not a joke consciousness is an immense light, like many lamp that create an immense light, any enlightened that annihilate into an

eternal resurrection added one more lamp to the eternal flow of consciousness awareness without disturb the other lamp, it just create more intense light with is essence data of intelligence fragrance quality , and the source is full of essence of many thousands of enlightened one who live flow eternally into the law of eternity, and actually constitute make up awareness and universal consciousness, they flow through intrinsic to the universal body and of course through universal consciousness, and that were you can meet them again and again, and certain they are source of immense intelligence you on earth are the receiver of the lamp, the legal representative ha ha you like it , basically the source of intelligence is consciousness an oceanic light consciousness, when you are link connected it happen simultaneous like internet, if you are link you have the connection it open up if you are not link connected the line is shut down , that the case of the unconscious dark persons , they are link connected to their mind unconscious and they think they are intelligent no is just information, you are loaded with thousands of program that you have down loaded from books television computer . I tell you a joke once in Delhi in an office of company of a factory industry was the officer there and suddenly someone nook at his door , the man to show that he was busy he take up the phone and acting like he was talking at the phone, because is good to show that you are always busy , and then he open the door and he was talking in the phone and he say ok for one million dollars ok no wait let make two millions dollars, at which the man who come in the office say but mister I am the one of the company of telephone, and I am just come in to say that the line of your phone is not connected some problem happen and it as to be fix up , that how all of you are, you are talking into a phone that is not connected, the bridge the link to consciousness is your inner being and until you live into the unconscious mind you are not connected link ,ok one million dollars no let make two millions dollars I would let make 3 million dollars ha ha ha . No sorry a little fun is always good, I was going to say something else the organic unity of a human being has seven astral body , the physical body the etheric body the astral body equivalent to the unconscious the mental body the spiritual body the Cosmic body and non being Body this the full rainbow color of the structure of your organic unity, I will not explain the full function working of each body and the methodology now is to long maybe in the next chapter, or you can look through my books I have write beautiful chapter on the seven astral body very accurate , but I will tell you a particular of the methodology how it unfold , so a man into the physical body is man into the etheric body is woman into the astral body equivalent to the unconscious is man into the mental body is woman, into the spiritual body neutral into the spiritual body the gender is transcend into the Cosmic body neutral but the size change into Cosmic size, into non being body neutral and the size vanish into infinity, no time no space no forms no duality into non being body they annihilate totally , and a woman into the physical body is woman into the etheric body is man into the astral body is woman again and into the mental body is man, and here is the difference into the spiritual body neutral same into the Cosmic body neutral and into non being body neutral beyond gender , this happen for a law of dialectic of complementary opposite that function every were in nature, also into the methodology of astral body , so a woman into the mental body is man and to transcend the male mind is a little beat more hard, instead a woman into the unconscious into the astral body is woman and that why woman are more sensitive intuitive receptive and that why woman are more easy to get enlightened because basically everybody lives unconscious, and to have an unconscious woman is a better shape to get enlightened instead man into the unconscious is man and is more rock like it repeal consciousness is more aggressive more violent not sensitive not intuitive because is unconscious is male , woman instead into the unconscious are woman a great advantage towards consciousness woman should seek enlightenment more through the unconscious because into the mental body are man more rock like , and man should seek enlightenment more through the mental body because into the mental body man is woman more receptive sensitive intuitive , so this advice to woman go for enlightenment more through the unconscious and this advice to man go for enlightenment more through the mental body this for a law of dialectic that function work everywhere in nature , and this probable is the motive why religions out of date expire spread lies about woman , because a woman is pure sensitiveness intuition she is woman into the unconscious and that is the ordinary dimension that humanity lives into the dark unconscious believe or not but it work function like this in fact a woman is more irrational then man the unconscious is more irrational a man is more rational because into the unconscious is man more rock like that is it this advice may help many seeker on the path .....Ok...Angelo Aulisa

The new book Mysticism & Physics Chapter 18 Angelo Aulisa

The melt down & new dawn of consciousness

Hi friends , look like that instead of a new dawn of civilization and consciousness is gone happen a melt down and climate change , I just see twice a documentary on the melt down of the ice shield on north pole and salt pole and I guess also the melt down of all glacial of the mountain , due the pollution of gas Serra and different kind of pollution that human being are creating blind and wild , the temperature on the planet is rising already since decades , The cause is the use of oil petroleum , and factory that produce gas Serra industry , cars flight , coal , but not only is suggest that all nuclear atomic hydrogen test of weapons rise tremendous the contamination pollution of gas Serra consequence the rise of temperature, and for sure the greatest cause of rising the pollution contamination and consequence the temperature, are the wars the bomb of all kind that they throw into the planet rise tremendous the temperature , now scientist make prediction that within two century the ice shield on north and salt pole will melt completely totally , but to me are very partial because we have no reference to calculate the speed of the rise of temperature and consequence melt down , because this so call civilization has develop more in the last hundred years that in the previous ten thousands of years , so we have no precedent to evaluate the melt down and rising of temperature, all the calculation are a little beat a guess work , for example they were saying twenty years ago that temperature will rise of 2 or 3 point in the next twenty years, but this is not true in Italy last summer for six months almost as being 40 point every single day that means that it was 10 point more of temperature every single day , and so all over the world in summer , and more the violent and catastrophic storms are already happen all over the world due the rising of the temperature of the water of the oceans , it maybe that the prediction are wrong and the process of melt down of the ice shield in north pole and south pole and mountain glacial is much more speed then thought , Also because the explosion demographic of population wild is running so speed , triplicate every 20 years that means if today we rare 8 billion in 20 , 25 years we will be 25 billion people on the planet, and everyone require transportation flight working in factory industry that create gas Serra and pollution of all kind , is natural consequence and also 25 billion people require an amount of energy incalculable that our planet cannot provide , and 25 billion people each one produce 3 tons of garbage to recycle that our planet cannot provide , the situation is dramatic I always through my books have talk warm about birth control for thirty years stop to birth to re equilibrium re balance the demographic explosion our planet cannot afford such a big population in terms of food , in terms of energy , in terms of space in terms of pollution gas Serra, and I want included the recycle of garbage that each one produce 3 tons every year we are going to be submerge by garbage, and become cannibals eat each other because there is not that much aliment food , and live in darkness because the world as very little energy , and this 25 billion people require transportation cars they are all small Gengiskan, that they want 3 cars 2 motor bike flight 2 time per months etc....The prediction of melting the ice shield in two century is wrong, right that English scientist that say by the end of this century it will be no more ice shield in the north and salt pole and no more mountain glacial , consequence the rising of the oceans of 200 meter and plus consequence many city nations will disappear forever , even if we build big huge dam you cannot fight against the oceans it will be such a huge violent storm, because the oceans they will be huge and warm that will trigger wind at 400 mile per hours devastating all and everything, the situation is dramatic , also because it seems that there is no possibility of changing the trend of this humanity mad neurotic running bling behind money and power prestige , careless of the catastrophe destruction of the planet that are creating . Climate change and melt down of the shields of ice and glacial is a very dangerous situation, is very important to rise consciousness about this subject urgent, to me the only possibility to stop this destruction are science and physic to give huge resource to physic to try to reduce pollution gas Serra through sophisticate equipment and to face this task square if is not too late of course. Together this humanity as to rise in consciousness awareness in civilization in intelligence, through meditation silence love, compassion understand, that the situation is grave and only a new dawn of meditation civilization intelligence consciousness awareness, can save the planet from this tremendous calamity , the answer is always a conscious aware world, through meditation, an individual become more sensitive more intelligent more lovely towards the planet earth the lotus paradise, and towards the organism of the universe and of course more conscious aware of where is living into an organism called earth and universe , and the individual as to play the major play role of being conscious to not polluted this earth to avoid pollution of all kind gas Serra plastic garbage etc... this humanity as to weak up from unconscious sleep hypnosis as soon as possible urgent, hypnosis means sleep induce , the individual as to weak up right now from the unconscious behavior and do is best play is part to stop this chaotic state of affair , starting from stop to birth for thirty years no birth , especially in Africa China India but all over the world basically, we get no more resource space food energy, cars that polluted the planet tremendous even if now they will start electric cars but they have battery that polluted even more , yes we recycle them I want see , is no more allowed to reproduce children blindly , the world is over crowed , and for thirty years is necessary a balance re equilibrium absolutely . For example all this problems of immigration in Africa Asia China you cannot go on reproduce ten children for family without have guaranty of economic support, and even if you have economic support is not allowed to reproduce children for 30 years the west country are collapsing, in Africa they are one billion and 250 million people the west country cannot go on guesting you blindly is natural is not racism , is just natural they them self-have no work no space no resources , but this is just to say that birth control absolute for thirty years is a priority of the world all over the world, to prevent climate change gas Serra rising of temperature, the demographic explosion is one of the main reason believe or not but is true, the great wise people that they go on saying it since 50 years and nobody hear them, enough birth control absolute is a priority of the world . Or otherwise the consequence are gone being dramatic as shown on the earth flooded documentary , thank you to the producer of such documentary who raise consciousness about a very urgent problem of the planet , I sustain that the answer is always a world meditative conscious aware, civilize intelligent can maybe stop this tragedy of climate change and melt down of ice shields in north south in the mountain glacial , the world by the end of this century is gone be submerge with the oceans water and I would like to rise a question they are many nuclear power station on see level, what will happen to the nuclear power station the will contaminate polluted all the oceans with radiation and radioactivity, so we will have huge oceans less land and the oceans will be radioactive with nuclear radiation is a big tragedy ..... this short chapter was obligatory to rise consciousness about a very urgent problem of humanity very urgent ....the melt down of ice shield and mountain glacial ...I love very much mountain glacial in my life I have make many track I love it them the glacial very much ..Angelo Aulisa

The new book Mysticism & Physics Chapter 19 Angelo Aulisa

The melt down & new dawn of consciousness part two

Hi friends , and so new born children are not send by God , because God do not exist is a disease called anthropomorphism the believe in God of this humanity , is an hypothesis at the most an hypothesis superstition lies , a neurosis of the ancient unconscious primitive barbarous humanity there is no God , and if it was one he look like that is a sadomasochist because many family in middle east India Africa China already have ten children, and no food no close no roof no money and God neither the less he send one more he is a sadomasochist that he like torture this humanity , no new born children we program them such event , and as I say in the latest chapter thirty years of absolute stop to birth of children is absolutely necessary , to avoid climate change , rising of temperature gas Serra pollution of all kind , this planet is over crowed it can afford one fourth of the population of today that means we are 8 billion try to think that in 25 years we will be 25 , because it triplicate every twenty five years , we will be submerge with garbage no energy no food it will be cannibalism eating each other , and the temperature rising will be a night mare the melt down of ice shield and mountain glacial will accelerate hundreds of time, and the immigration situation will be a night mare , the west country are going through this night mare specially Italy that is the gate from Africa to Europe , you know the African are one billion and 250 million if you live the gate open like Italian blindly are doing within twenty years Italy will be a colony of Africa , is not racism but the situation cannot be sustain any longer, Italian they should reclaim their borders , they have no work no resources no spaces for them self . Anyhow the birth control stop for 30 years is a priority of the world to avoid the calamity of melt down of ice shield in north and south pole and the melt down of ice glacial of the mountain , and to avoid calamity of climate change like huge terrible storms and rising of temperature everywhere , birth control stop of birth is the main cause of this calamity no more blind production of children , and basic is a big responsible you need economic support combining of genes , and conscious compatibility , enough of blind wild procreation all over the world specially in the sensitive country like middle east they have family of 50 children because of multiple wife , and Africa they family of dozens because of absence of intelligence , and India they have family of twenty children because of growing there number of supporting there religions , and China were they are one billion and 300 million because they have to be the super power of the world and basic all over the world stop birth at zero rate to save the planet for tremendous horrible terrible calamity , weak up from unconscious sleep hypnosis of religions old and out of date expire superstition pseudo that they preach lies and lies from the begin of their religion to the end fancy tale lies a bunch of neurosis traditions and lies , a night mare of the past ancient humanity , you know if you don't weak up to abolish this fanatic pseudo false absurd superstition religions they will destroy this planet in the next hundred years, with conflict wars of different philosophy culture futile things, and preaching blind procreation because it pump up the ego of the number of followers of their religion , they are senile mad insane neurotic please abolish ban stop all of them the old ancient pseudo superstition hypothetical religions now, tell to all of them to resign now or they will destroy the planet with their ignorance blind tradition of the past dead, and a bunch of neurosis infantile childish absurd fancy tale for retarded children . The religions expire and out of date they play a major destructive role in climate change, rising of temperature melt down of ice shield north and south pole and melt down of ice glacial , with their blind tradition of the past that children are sent by an hypothetical God, and with conflict wars constant since thousands of years this wars for futile motivation, not only create massacre and bloodshed killing and destruction, but they rise the temperature of the planet with throwing of bomb of all kind and polluted contaminate the planet irreversible, is time for all the old and out of date expire religions to give up resign before that is to late enough, what do you want physic science mysticism have demystify all them are a laughing stock for retarded crippled children so kind give up we are running out of time free humanity from your stupidity ignorance senile and insane free humanity from your enslaving people with absurd out of date tradition a bunch of neurosis , the greatest dangerous for humanity are the out of date expire religions they are anti-life anti celebration against humanity intelligence against science against physic against freedom and civilization , they are tribal sectarian each religion is a big sect , right Gurdjeff who was saying this hundreds of years ago the religions are the greatest dangerous for Humanity and against life and celebration and freedom and civilization against intelligence against truth , against physic and science and against mysticism the science of inner mystery of human being , enough is enough resign at once all together the old out of date religions expire and free the world from this dark unconscious hypnosis, before that is too late , the world will be lead objective for 90 per cent by physic science, with huge resources to restore eco system climate change melt down of ice shield and glacial of mountain, and basic to create a new world at measure of the new men meditative intelligent conscious aware, of where is living into an organism of planet earth and universe , science and physic will change the face of our objective planet and mysticism the inner science of men will lead the inner reality of men with meditation basic you close your eyes and bring the attention in the within dimension of vertical dimension of deeper and higher, the present is the gate less gate to your inner being the gravitational energy field call dark matter will bend the light consciousness deep within your inner being and get centered rooted at the deepest point of your inner being , you are not your body not your thought you are not your emotion not your unconscious staff you are not your senses but you are just a pure crystal clear witness consciousness, and an empty mirror that what is your soul an empty mirror that reflect all and everything in the moment next moment the mirror is empty, is not coolness but just the inner nature of your soul and then a pin drop silence is the gate less gate to your inner consciousness, the expansive energy field call dark energy will rise up your light consciousness at the horizon of event of the universe expanding it your light empty consciousness on and on with the same pass of the universe is a break through divine sacred holy, your size change into universal size at this point your inner being is the whole universal body is divine sacred holy, you in essence are intrinsic to all living being and forms of the universal body, you are the universe itself and on deeper with meditation emptiness nothingness are the gate less gate to non being body, were time space forms duality of mind totally annihilate and consciousness also annihilate consciousness is always in relation of a subject or object, into non being body no subject or object survive so consciousness annihilate into formless awareness unfocused and relation less, this awareness is just an I am ness a great relaxation, and ultimate essence of non being body of the law of eternity the size at this point vanish into infinity into eternity begin less endless never born or die, that the meaning of eternity and core of the mystery of the universe and of life and death and of all duality , the ultimate canvas were the universe is display paint , above beyond bigger and vaster transcendental to the universal body itself , an open relativity infinite, freedom sacred holy divine a pin drop silence subtle ecstasy were the immortality and resurrection reside and unfold , the resurrection is a conscious alchemy not gross not material not physical, from unconscious to consciousness awareness to into eternity itself, into non being incorporeal body called it , an here you in essence are at home resurrected awake from unconscious sleep and various hypnosis , here you are free in freedom each individual in essence is eternity itself one indivisible, one eternal immortal resurrect free in freedom forever in peace forever enlightened, forever enlightenment is your birth right and your true inner nature, you have always being enlightened is a matter of remembrance of rediscover it within your inner reality, and then the new dawn of consciousness awareness civilization intelligence, the ultimate convergence of evolution quantum leap that our DNA aspire for immortality resurrection has happen, the triumph of civilization the accomplishment of human being on earth the ultimate refinement of consciousness has happen, you in essence return at home like zero like a circle of zero, the journey is from eternity to eternity from here to here , when you leave the body for an eternal resurrection you are not going anywhere, is from here to here if it is a real resurrection of course, if is a changing of house body a reincarnation then you go for a Cosmic journey within days months you will have a new body a new opportunity to accomplish the full journey, of your refining your consciousness into light consciousness into enlightenment into bliss into ecstasy into sacred divine holy peace bliss ......Angelo Aulisa

The new book Mysticism & Physics Chapter 20 Angelo Aulisa

Doing things together

Hi friends , I love Obama the best president that unite state as ever had , the most intelligent the one that as change the face and the attitude of unite state , he has done great things pulling out almost all the military from everywhere , and give a face an image of unite state human democratic of freedom , some flows were there the behavior against Israel and other small things but basic the best president of America he try to abolish the weapons sale on the nation of unite sate saying that was insane to go around the country with a weapons , and he as dream of a world without nuclear weapons atomic hydrogen weapons a beautiful dream , he try to pierce this dream of a world without deadly weapons as much he can , a very sensitive intelligent civilize individual of the world thank you again and again thank you, he try to shut down torture prisons , and he try to give to the Americans a free health care as much he can ,very important this subject all over Europe the health care is free own pay by govern a very civilize being , and finally he try to bring the world together like Cuba , Iran , and other place Asia etc... a great men really creative positive life affirmative , in the case of Cuba and Iran he under estimate the individual and collective unconscious of that nations but probable they was no other option of doing it that way , because the individual and collective unconscious of that nations will in the future assert itself , but they was no other way to do it a great creative men and very intelligent and sensitive , In Syria he support the movement of freedom and democracy thank you very much , he try to create as many bridge he can with other nations even if they were standing in different vision and understand great this is my respect and gratitude for intelligence thank you Obama thank you . About the actual situation of the world which is very tense and dangerous , I like the formal defense minister of America , I propose this diplomatic option , the record nice of North Korea , and Iran as nuclear power nations and to welcome them in the club of nuclear power nations , they have acquire this nuclear deterrent I don't care how but they have so is better to record nice them booth and invite them in the club of nuclear power nations , they will have to sign subscribe peace deal with the all other nations , and the lift of all sanction against those two nations so that they can reintegrate with the rest of the world , is better to have those two nations North Korea and Iran record nice officially, that to have them on the lose wild with sanction , and the next step is to call a worldwide meeting of all nuclear power nations a peace table , and welcome this two nations North Korea and Iran and tell them we were waiting for you to complete the picture of all the nations who hold nuclear atomic hydrogen weapons deterrent , you were thinking that you have acquire something special but is not so , we are talking how to eliminate destroy deactivate all the nuclear weapons atomic and hydrogen weapons since years , and we were waiting for you booth to keep on the debate and peace table , we all the nations who hold such a deadly power and device we are talking since years how to destroy eliminate deactivate all of them and you booth Iran and North Korea are invite to the conversation debate of disarmament of how to eliminate deactivate all of them , and we have reach to this conclusion that the elimination destruction deactivation of all nuclear atomic hydrogen weapons can be done only all together, all the nations that hold such a deadly power that can destroy vaporize the world in hours , together simultaneous we begin the process of deactivation elimination destruction of such a weapons , together simultaneous because is the only way to trust each other, that all of us nations who hold such power deadly we are sign and subscribe a deal of disarmament and elimination deactivation and destruction of all this deadly weapons called nuclear atomic and hydrogen and chemical , here on the table the deal of disarmament, but we were waiting for you booth north Korea and Iran to complete the picture of the disgrace nations who hold such a deadly destructive dangerous like hell power for the all world , please sign and subscribe the disarmament deal as we the nations that we hold the same disgrace dangerous deadly power we have already sign and subscribe it , once you both sign tomorrow at 12 AM it will begin the process of disarmament deactivation destruction of all nuclear atomic hydrogen chemical weapons of the world, and it will begin simultaneous in all the country of the world and you both are welcome to join the deal of disarmament deactivation destruction of all deadly dangerous weapons , so then that the world in the future can live safe in peace love compassion intelligence in consciousness awareness , and the relation between nations in the future will be through measure our intelligence consciousness awareness, and not through measure the weapons that one has , that was old out of date expire primitive barbarous unconscious way of relating inspire by the ancient out of date expire tradition of the past the tradition of wars of the past is abolish forever and ever, the new UN the new world relate between nation through intelligence love peace compassion awareness and consciousness , and this first step of abolish ban stop all nuclear atomic hydrogen chemical weapons and conventional weapons too is the first principle of a new civilization , no weapons no military no wars no factory who produce weapons anymore is a crime against humanity to build device of death call weapons, who can kill an extension of your own consciousness in the forms we are 8 billion in the formless consciousness one oceanic consciousness a oneness a mystical union of one empty consciousness , extension of one another sacred holy divine, each living being and forms of life is sacred holy divine and is a crime against humanity to hurt interfere kill an extension of your own consciousness , hence all trade of weapons is abolish ban stop at once , nuclear atomic hydrogen chemical weapons and conventional weapons and military wars , and consequence bloodshed massacre death destruction kill is abolish forever and ever , A new era of civilization and the first principle of civilization is a world with no weapons, because love consciousness awareness meditation, will be the tread that bond and relate all world together no weapons are needed anymore , And so here on the table the deal of disarmament and peace of the entire world . You like it a dream to sign and subscribe right now. Today at euro news the journalist say that the world in the last 5 years as invest 500 billion on nuclear weapons is a lucrative market , they can invert such investment in beneficial item for humanity like sophisticate equipment to reduce gas Serra pollution, that create climate change and the melt down of ice shield on the poles, and the melt down of glacial in the mountain , they change their production in item beneficial for humanity and we can use those resource immense to reduce starvation around the world, misery around the world if you take the all resource of building weapons of all kind of all military of all wars and invert those resource transform those resource in beneficial item for humanity we can stop climate change melt down of ice shields on the pole and the melt down of mountain glacial and put an end to starvation and misery at once, all over the world and rebuild a world biological city at measure of the new men meditative civilize intelligent conscious aware, we will have this world the lotus paradise in peace love compassion harmony balance celebration playfulness, and creativity of all kind will flower like lotuses, and this body the consciousness awareness a new era of authentic living of real living, probable for most of humanity the first time of living in meditative conscious aware way for the first time will test what real living in harmony balance mystical union with the universal body and universal consciousness, and awareness into non being and eternity means, all together another living another plane another dimension less dimension, into eternity begin less endless never born or die that the meaning of eternity. and core and source of the mystery of the universe and of life and death and of all duality , ultimate canvas of the universe were the universe itself is display paint , you should try this true real authentic living, is a miracle in itself before that you destroy the all planet test the nectar that life is for real, and you will change your destructive attitude into creativity into love into harmony and balance, with the whole universal body and eternity itself, human being in essence is eternity itself never born or die never begin or end immortal and resurrected, already within is inner reality is matter of remembrance rediscover and not discover , rediscover is already there within you , you have only totally forget due of living in extroversion , a little balance of introversion and extroversion of thesis introversion and antithesis extroversion and synthesis of your own inner being, and from your inner being simultaneous you are bridge link connected with the all existence , with consciousness awareness into non being eternity itself, one intrinsic indivisible, enlightened forever and ever , and then you can celebrate share your inner love and have a cup of tea ....Angelo Aulisa

Published on October 20, 2017 04:13 • 46 views

The new book Mysticism & Physics Chapter 21 Angelo Aulisa

Doing things together part two

Hi friends, Einstein at his own age many years ago was asked , what will happen if a third world war happen , he answer short he say it will be no a fourth world war , it will be the end of this humanity as we know totally vaporize vanish , and it was many years ago , try to think today were the nuclear atomic hydrogen chemical weapons are thousands of time more strong and deadly and thousands of time more as quantity , they are nuclear weapons to destroy twenty thousand time the world , and if you put together atomic hydrogen and chemical weapons we can destroy vaporize the world 50000 time, and if you added all conventional weapons 60000 time the world can be vaporize totally destroy totally 60000 time that means that each human being can be killed 60000 time , is not this neurosis madness crazy , in sort pathologic sick insane , the old little insane men of today as cross all red line all boundary of madness and seriousness , because only a very serious men that live in is mind unconscious totally identify with is insane thought and emotion and tradition of the past, dead and out of date can reach a point to dig a big graveyard for all world as the old serious sick little men as done , but how far can go the stupidity insanity of men you get nothing else to do that build device of death called weapons, that if an insane men with abnormal ego unconscious push the button of one of those only one missile can kill ten millions people in a city , I ask myself how vicious insane crazy mad unconscious human being have become , look we walking in a razor edge is too dangerous to keep on this vicious game that nations and politicians are doing a wrong step on the razor edge, and an irreversible tragedy massacred bloodshed disaster will happen this time will create an havoc into humanity, you are not playing with toys this weapons nuclear atomic hydrogen chemical are dangerous like hell, and they can put an end to our green planet and to this humanity in hours this humanity and planet will be vaporize in a cloud of smoke , please to all politicians govern come to your sense and abolish ban stop all this deadly weapons from nuclear atomic hydrogen chemical and conventional too, deactivate eliminate destroy all of them abolish ban stop all military and wars at the same time, all of this is a big crime against humanity and against all living being and against all intelligence evolution evolve in thousands of years with heavy sacrifice by this humanity , come to your sense and come together in table of peace as I say in the previous chapter, and abolish ban stop deactivate destroy eliminate this dangerous treat menace that is over standing humanity, you have gone too far is time to come to your sense and begin table of peace of disarmament of all this hell that unconscious insane senile you all have create , as I say we are walking in a razor edge a total destruction can take place if humanity fall in the wrong side of the edge and don't listen to the intelligence of the world , but a new dawn of civilization intelligence consciousness awareness, can happen too if humanity is listening to the enlightened awakened one to the Buddha and fall into the right side of the edge a total transformation, just listen to your own inner being and consciousness awareness, and come to your sense and deactivate destroy eliminate all nuclear atomic hydrogen chemical conventional all hell that you have create unconscious asleep, hypnotize by the collective unconscious were all past tradition dead and out of date are store , Open your eye weak up a third world war will have no winner, all will be vaporize destroy finally and forever and they will be no winner and maybe this is the motive why is not yet happen, they will be no winner all losers humanity will be lost forever this planet will be lost forever totally vaporize, it will be left a desert of radiation radioactivity for hundreds of years to go , hope the intelligence an common senses prevail , I insist because I am not ready for a global suicide, destruction vaporization of this humanity and planet , as the individual unconscious and collective unconscious of humanity is pushing for , the collective unconscious of this humanity is saturate to an extreme point, and the collective and individual unconscious wen is saturated to such extreme point and full of stem , it create a within cloud call Tanatos that as suicidal tendency that why many people commit suicide due this within cloud into the unconscious that Freud itself called Thanatos , is like men as a limitation of afford suffering and misery anguish frustration, then he give up and many commit suicide , same in large scale the collective unconscious of humanity it build up within the collective unconscious a cloud of thanatos of suicidal tendency , because all as fail democracy as fail, religions have fail, society have fail system have fail the family as fail , life itself as being a failure in everything you have being doing, hence the tendency to a global suicide basically human being men as never live the life and this hurt like hell this is the within dynamic of the collective and individual unconscious, that create the cloud call thanatos that trigger individual and collective global suicide , but I am not ready for a collective global suicide as I am into the world and if you do this crazy act of a third world war I am included in it I insist in give you light understand , that a new era is at the gate a new dawn of civilization consciousness awareness, intelligence is urgent needed and possible, a new way of living at a different plane that you all have never experience never , and is a divine sacred holy blissful ecstatic way, meditation silence love the main path to this new way of living, to the nectar of life to the miracle that real life is , I am ready for a new dawn of civilization meditation consciousness awareness, playfulness celebration love, peace dance music joy , and with me all the community of Zen that I belong too , so we are ready for a new dawn of civilization life love laughter ,meditation dance celebration, rejoice into life abundant , that why my insistence to protect myself and my community of Zen awakened one that I belong to, this explanation to sheer light on this delicate subject was extreme important for all of us the world that I belong too , because we are the new men the salt of the earth, the intelligence of the earth and we have to keep on the dream of our beloved master at any cost , please come to your sense life is beautiful, a miracle to love, destroy deactivate eliminate all the nuclear hydrogen atomic chemical weapons and conventional weapons too, we love life and all of you are welcome to join our community of peaceful joyful lovely individual, and enlightenment will be just a gift for all of you I guaranty , I am here were iam but my future my life my vision for life is the Zen community that soon I will be back this time forever and ever .....Angelo Aulisa

Published on October 20, 2017 10:27 • 42 views

The new book Mysticism & Physics Chapter 22 Angelo Aulisa

The seven astral body of an organic unity

Hi friends , finally I finish all my suggestion at the moment for the contemporary world , at the moment because is an going process that never end but I am happy that I can go on write my book on the things I love on the subject that I love and I am a specialist about mysticism , science and physic , I would like once again to spell the seven astral body of an organic unity I have done this in many of my books very accurate , but once more because I think that if humanity understand the importance of the seven astral body of an organic unity it will be more easy for humanity to drop abolish completely all of the old out of date religions superstition pseudo a bunch of neurosis a night mare of the past ancient humanity unconscious ignorant and primitive barbarous , basically the world is divide in big sect call religions , hypothetical and hypocrite and lies from the begin to the end that what religions are lies and lies hypocrisy and hypocrisy hypothesis and hypothetical, about the fiction of God and they generate a disease called anthropomorphism the personalize of God, that make human being crippled conscious retarded . So is important the resignation abolish of all religions, and to come in give to humanity scientific approach to reality to existence to life and death, the seven astral body is a scientific factual method device to approach reality in an authentic way universal and eternal. Human being is not mere matter or just physical objective , if it was so life would have being like Sartre say meaningless a story told by an idiot full of fury noise signify nothing at all , and for many people in the world it is so 99 per cent of people in the world live like this only in the objective material dimension only matter and physical , money power prestige , ego mind unconscious horizontal from abs , xvz , at infinite regression gazing the horizon tomorrow is gone be love tomorrow is gone be better tomorrow is gone be the happiness , but what come tomorrow is death , because in life everything change, life is changing every split second the only thing that do not change is death , and death will come even if to you 80 , 90 years look like a long time in a space less space timeless time of eternity is just the blink of the eyes and is gone . Life is much more then matter physical , life is an infinite mysterious journey into the mystery of your inner reality into the mystery of the universe into the mystery of eternity is an eternal journey that end nowhere and begin nowhere , but to have this quantum leap into the dimension of being and consequence into the vertical dimension of deeper and higher of here and flowing into consciousness moment by moment meditation is a basic requirement or silence or love or dance or painting or laughing or sculpting or singing or whatever is your attitude affinity to get into a mindless dimension of no mind , Meditation is the easy way path method , You close your eyes bring the attention within yourself the present is the gate less gate to your inner being, the gravitational energy field call dark matter will bend your light consciousness deep within your inner being, you relax and let go within at the dippiest point of your inner being a pin drop silence peace sacred holy divine, is the gate less gate to your inner universal consciousness the expansive energy field call dark energy will rise pull up wards your light consciousness up to the event of horizon of the universe on and on in synchronicity with the universe, is a sacred break through divine majestic you in essence flow intrinsic to all forms of the universe and living being your size change into universal size anatta your inner being annihilate into the universal body and at this phase stage your inner being is the whole universe itself , and at this stage the flowing energy field call dark flux will flow move circulate your light consciousness in a Cosmic play irrational in and out of the core and source of the mystery of the universe the law of eternity, in a festival of light consciousness circulating moving flowing in and out of your inner being which at this stage phase is the universal body itself is divine sacred holy an ecstasy a bliss, and on and on nothingness the absence of things emptiness is the gate less gate into non being body, were time space forms duality of mind totally annihilate and consciousness too annihilate because exist always in relation to a subject or object into non being body no object or subject survive so consciousness annihilate into formless awareness unfocused and relation less awareness, is just an I am ness a great relaxation let go into non being body incorporeal body everywhere and nowhere in particular expand all over ultimate essence of the law of eternity begin less endless never born or die, that the meaning of eternity an infinite open relativity an infinite freedom infinite bliss infinite peace infinite silence intrinsic to silence a subtle ecstasy that surpass all understand , here the size vanish into infinity and here you in essence are at home were immortality resurrection reside take place and unfold , the resurrection is a conscious alchemy from unconscious to consciousness awareness into non being incorporeal body into the law of eternity core and source of the mystery of the universe and of life and death and of all duality, ultimate canvas of the universe bigger and vaster above beyond transcendental were the universe is display and paint , here the unconscious sleep and various hypnosis totally gone here you are awake aware and enlightened forever and ever , enlightenment as happen finally and forever and ever . I start the introduction of the seven astral body like this from very far away because then it will be easy for you to understand the methodology of the seven astral body which is a very powerful authentic path way to approach reality scientific and factual , look if you understand the methodology of the seven astral body your life will change transformed into bliss playfulness celebration into a peace endless, life will become just a beautiful Cosmic play it will change completely your plane of living and your gestalt focused , so I want approach the seven astral body in very deep way , because humanity up to the mental body as reach more or less the intelligence part of humanity as explore up to the mental body , but they are totally unaware unknown of the conscious body or called it the spiritual body and the Cosmic body and non being body , and this last 3 body are the real quantum leap that humanity should got assimilate and known pierce totally , so then a convergence of evolution happen into the DNA of each individual a mutation that will make his life a bliss a Cosmic play , a celebration a peace sacred divine holy , believe or not but this methodology that I am going to explain express unfold once again is very powerful tool of transformation very effective and scientific factual , so I have start from very far away and now I am going to unfold it to you all follow next chapter part two here end part one ...... Angelo Aulisa

The new book Mysticism & Physics Chapter 23 Angelo Aulisa

The seven astral body of an organic unity part two

Hi friends , so the full structure of an organic unity as intrinsic to it seven astral body , physical , etheric body , astral body , mental body , conscious body (spiritual body in old terminology) Cosmic body and non being body , The major part of humanity knows and live at the physical body (Donald Trump for example knows and live at the physical body I joke ha ha h) but it is so the physical body move in time ten clock am is ten am and it move in time , the etheric body which is the emotional layer of your organic unity can move in space for example if I am here ten am talking to you but you are thinking of your mother or beloved in New York you are not here with me but you have move your thought and emotion in New York with your beloved or mother and I am doing a monolog talking to myself because you are not here present to present with me but you have move in space but ten am remain ten am , you have move in space but the time remain the same , and it work function perfect like this because therapist use to make reiki session at distance to heal the patient friend that they give the heal session by appointment for example , so the etheric body move in space time remain the same , beautiful you can move all over the world universe you just bring your attention were ever you want basically you are were your attention his true is like this , your emotion can move your thought can move in space , then we got the astral body which is your unconscious since you born up to the present moment of your life all is record into the astral body all event of your life wounds trauma conditioning repression beautiful and bad event in digital color , in the astral body you can move back words up to the primal scream , and readjust revive relive all the event of your life , but in the astral body you can move only back words up to the primal scream , is very useful because you can heal bring light into your dark unconscious and burn the wounds the conditioning the trauma literally burn them off you just have to witness them with your witness consciousness and is the fire that burn them off they are no medicine or tablets for the unconscious the medicine tablet is to bring light in witness with your witness consciousness all of them the observation witness of the wounds trauma event evaporate burn the wounds trauma off the Cosmos reabsorb them the witness observation of the unconscious is the healing the medicine that burn them off , many they use the psychoanalysis session the doctor to get heal the unconscious to burn the wounds to get free of trauma , but if you are centered into your inner being and witness consciousness you can be the psychoanalyst of yourself and basic is better because alone in intimacy you can unfold the most intimate wounds trauma that you would not do in the psychoanalysis session with the doctor, because the intimacy is lost , so the astral body is equivalent to the unconscious , then we got the mental body which you can pierce move back words up to your birth , or forward into the future imagination but one at the time , the mental body is made of factual memory of event of your life more on the surface and not insight event of your life , but is easy to mistake with the insight of the unconscious , and is concern with the future when you move forward with imagination but you can have also revelation of future always through imagination or thinking , very important the mental body is huge almost infinite the brain is an organ which humanity still today knows very little it as infinite intrinsic quality and possibility , for example telepathy , revelation of future , prediction of future , like Nostradamus , the dimension of dreaming which is related to the unconscious but dreaming can have many layer also towards the past and future, you can fall easy into past life dream when you dream or you can fall into future event of life and have revelation but that is rare and anyhow when you dream you should be awake as a witness consciousness to remember the dream , but if you are awake as a witness consciousness into the dream it broken and you awake , if you are really awake as a witness consciousness dream vanish don't happen you simple sleep blissful without dream , but many people unconscious can remember the dream they make like my mother and they can tell you the dream with such a clear detail incredible but all unconscious , because many time I tell to my mother remember you are not the dream but the witness consciousness that witness watch the dream , and dream they can be of thousand nature , so get centered on the seer the witness of the dream because that you are that is your soul your witness consciousness and she got it very clear , if you are awake as witness consciousness even into sleep dream don't happen take place your sleep will be a blissful divine silent sacred sleep all 8 hours and you regenerate rejuvenate completely during sleep you relax into the core and source of the mystery of the universe and of life and death and of all duality into eternity and you recharge your battery , if you dream is because you are unconscious and six hours are of dream and two hours are of deep sleep in those two hours you regenerate and recharge your battery , but for six hours you were dreaming constant , Anyhow the mental body is huge vast immense many people think that this is all that there is and stop there they think that more is not possible , but is just a phase a of the infinite inner journey you have to be very alert to move on and don't stop there because the mental body can deceive you very easy is very psychedelic huge that you may think that more is not possible , humanity up to the mental body as explore and knows this first 4 body and record nice them very well , but the journey as just begin ...here end part two of the seven astral body follow more in the next chapter ....Angelo Aulisa

The new book Mysticism & Physics Chapter 24 Angelo Aulisa

The seven astral body of an organic unity part 3

Hi friends , before to go on with the methodology on higher body a few accuracy about the physical body , wen I say that the physical body can move in time if is ten am it can move in time literally , that is understood that he can walk in fact you can walk from your home to your office and when you reach your office it must be 10, 15 am , you have move in time or you can take your car and go to town when you reach town it must be 10.30 am , that it do not means in the methodology that you have move in space , in the methodology to move in space means that just seating in meditation without moving a centimeter you move your etheric body to New York , or you move into your astral body at the primal scream backwards without moving a centimeter you journey within your astral body , same into the mental body backwards and forwards you pierce move within one at the time without moving a centimeter , is a conscious alchemy (spiritual in old terms but I hate the terms spiritual the new terms for spiritual is conscious) conscious alchemy not physical not gross not material but conscious , this to clear up some miss understand . More the mental body is a journey into your brain which is a vast huge infinite journey , in the recent years we have get to know a little more about the brain and how it function the neurons vector of chain reaction into your brain of thought emotion believe have being feed by the society system old out of date religions and are loaded with pseudo lies falsity and conditioning repression of all kind wrong notion , and that is why the human being of today look like neurotic it is full of neurosis , and when the foundation of your house are made of lies wrong notion conditioning repression falsity , soon or later the house collapsed and the outcome is altzaimer and other disease related to your brain living your full life in lies wrong notion falsity create a side effect of hypocrisy falsity that on the ultimate run of your life bubble up and disease as cancer altzaimer etc.. is the outcome , because you have live a full life in lies hypocrisy wrong notion that have completely take away your true nature of freedom spontaneity naturalness and finally the result show come out in disease , your neurons loaded are equivalent to your unconscious the astral body totally saturated loaded with lies falsity hypocrisy conditioning repression inferiority complex and guilty complex wrong notion in the end it collapsed the truth assert itself the coming out is serious disease , and also the premature deterioration of the cell of learning into your DNA is a consequence of the garbage store into your neurons , and when you cannot learn anymore is an early death , the brain function as a full network of synopsis too tissue , today we know a little more about the brain but still remain unknown at large , for example the most genius brain of the earth like Einstein it uses is brain 14 per cent , and he was really a genius , the average of quotient intelligent of humanity is that uses their brain 7 per cent , and the average of age of even an Einstein is 12 years of age and the average of age of ordinary humanity is of 7 years of age , childish infantile , Freud say that a man it do not grow anymore after the age of seven , because the foundation of his unconscious have being laying and he stop to grow , all this to say that the brain remain an unknown organ even today 85 per cent of the brain is unknown and not used , unless you are enlightened then you live beyond above the brain and your neurons are free of conditioning repression lies wrong notion wounds and trauma and free of inferiority complex and guilty complex , in short the neurons of an enlightened are free of garbage empty and they function as vector of consciousness which intrinsic is full of Dharma like love intelligence compassion peace silence, intrinsic subtle ecstasy celebration and playfulness an enlightened man is never ever serious for him is a Cosmic play all the time about anything , even death , then the foundation of your house the roots of your trees are in consciousness awareness in non being body into the core and source of the mystery into the law of eternity and you will be an healthy being, your house will never collapsed fall apart , you are rooted in your true nature of your inner being , in consciousness awareness and bliss peace is the outcome the consequence the result , all of this to say that the mental body is a huge infinite body that can deceive you to think that more is not possible and you may stop at this phase of the journey , but the real inner journey is not even begin , you are still on the surface on the horizontal plane from abs to vs z. at infinite regress in a vicious circle of mind ego unconscious on the dimension of having material and gazing at the horizon that tomorrow the real thing is gone to happen tomorrow the love tomorrow the happiness and tomorrow never come is death because death is the only reality that never change and finally death will come , otherwise life is a constant change mutation every split second ,what come is today now moment by moment flowing in consciousness awareness and got centered into this split second of the present gate less gate to your inner being and to the vertical dimension of deeper and higher, and flow into consciousness moment by moment into the present witness whole and everything from your witness consciousness, and reflect whole and everything like an empty mirror every situation that surround you reflect like an empty mirror and the next situation the mirror is empty again ready to reflect what come next, because that is the true nature of your soul of your inner being witness consciousness and be an empty mirror your soul is an empty mirror the chief quality of your soul , my mother understand it very well she was a Sanyasy the best sanyasy and she was not so much cultured that was the reason why was so easy to make her understand all inner dynamic of an organic unity she had no much ego no much knowledge no much ego unconscious mind she was simple and innocent , and as being easy to lead her into an eternal resurrection into Samadhi in peace and it as happen the greatest resurrection ever witness by me , now I say about my mother because if this enlightenment can happen to her I think that for you is gone be even more easy , enlightenment is the human right of every human being and is already the case within your inner reality is just a deep remembrance rediscovery and not discovery , and enlightenment means mystical union oneness with the source and core of the mystery of the universe and of life and death and of all duality with non being incorporeal body with the law of eternity a human being in essence is eternity itself, one indivisible never born or die no begin or end , immortal already resurrect just a deep remembrance 24 hours a day, try to think if you have a deep remembrance of yourself 24 hours a day is done is so easy you in essence are always at home centered into your inner being and consciousness , you know Battista the master of Jesus he stay in the Jordan river day in and day out and shout loud return , return to your inner being (soul) that the king is at hand is coming , he was saying exact this remembrance of your inner being 24 hours and not repent as Christianity give this wrong interpretation of repent is not repent in Aramaic was the right interpretation was return to you true nature of your inner being soul and all is done, is this remembrance that I am telling you , enlightenment is already within you just remember it rediscover it , my father which was in will chair because of an accident he could not talk or walk properly he understand this remembrance rediscover totally and I tell him the all story of Battista shouting in the Jordan river and he laugh and he got it very clear and he was grateful to me so much , that is this remembrance my father in the will chair got you can got all of you too easy is the easy method of getting centered into your inner being .. Ok here end part 3 of the seven astral body follow more in the next chapter the higher body.....Angelo Aulisa

Published on October 23, 2017 01:11 • 45 views

The new book Mysticism & Physics Chapter 25 Angelo Aulisa

The seven astral body of an organic unity part 4

Hi friends , it would be much more to say about the mental body but the basic important things have being say , we can go forward into the conscious body (spiritual body but the new term for spiritual is conscious me I spell conscious body , because spiritual is the outcome of a short time flirt of Marry , the holy spirit was a young beautiful Israeli man , and it look stupid to call something so sacred and holy as a conscious stage spirit, I refuse it so well know that conscious is the new term for spiritual) so the conscious body first of all into the conscious body gender is transcended, the conscious body is neutral to gender conscious is neutral to gender , the conscious body is a quantum leap into the vertical dimension of deeper and higher, the present split second is the gate less gate to your inner being , into the conscious body you can move backwards in time up to the super conscious up to non being incorporeal body , the journey is through your individual unconscious and collective unconscious and backwards into through the cosmic unconscious , The Cosmic unconscious is not record nice and known by humanity, I have try hard through my books I talk and explain allot about the cosmic unconscious, but so far humanity as not record nice or known the cosmic unconscious , the individual unconscious is huge immense an iceberg , the collective unconscious is huge infinite like the Himalaya mountain , infinite you can never come to the button of it , but the Cosmic unconscious is absolutely huge big immense many layers at least 6 immense infinite layers , the first layer is all the cycle of your past life thousands plural you go through all of your past life , (now one thing is important to say you can go through this inner journey with your witness consciousness awake and conscious then you will remember and witness conscious all of your past life, or many or you can go through this journey unconscious then you will not remember and the witness consciousness will be asleep , so it will be all unconscious as happen in the case of coma when someone go into a coma because of injure , then you will have a simple sensation of being reborn when you come back from the coma of new begin fresh ,) but if the inner journey happen conscious awake you will have a crystal clear remembrance of the all inner journey not only of past life but of the all layers which I say they are at least six big huge immense layers, and when you return into the body a new presence totally enlightened forever and ever , The first layer of the cosmic unconscious is overlapping the collective unconscious, are all of your past life that they remain in print into the cosmic unconscious in digital color as factual memory all action and things that you do remain record into the cosmic unconscious forever, until wen if it ever happen you journey into it in a retrospective journey backwards , then your journey into it and your witnessing of it of all the event of your past life is the fire that burn them off the cosmos reabsorb them all through your witnessing of them , and then it will be left a pillar of light empty consciousness, this process happen in small scale also into the individual unconscious and collective unconscious , your witness and observation of wounds trauma conditioning is the medicine that burn them of and you are heal the cosmos reabsorb them , and you are left within your unconscious heal and with a pillar of light empty consciousness free of wounds and conditioning trauma , Same happen into the cosmic unconscious when you witness your past life they burn off and disappear and you are free heal of all the past life, that you witness because the final heal is to have an unconscious free and feel with a pillar of light empty consciousness , Is a little beat like wen in physic you observe a particle and the particle through your observation change behavior or even disappear and reappear again or even the particle collapsed through your interference of observation , same happen into the unconscious when you bring light in witness consciousness in is the fire the medicine that burn off all the wounds trauma conditioning repression inferiority complex and guilty complex , and your unconscious get heal free feel with a pillar of light empty consciousness , because you have to know that in your present life many of the disease neurosis that you have and you are crippled are coming from either from your individual unconscious or from your collective unconscious but most many of the neurosis that make your life crippled are coming from your past life event wounds trauma conditioning , for example in your past life you may have being a prostitute or a thief or a murder or tyrannical Mass murder , for example and it affect your present life in very mysterious way , it crippled your present life without that you even know unconscious process , for example if you are a man in your present life in your past life you were a woman and if you are a woman in your past life you were a man for a law of dialectic that function hundred per cent in the evolution of your consciousness from unconscious , to witness and observe your unconscious either individual or collective or cosmic unconscious of past life can reveal to you the cause of many disease neurosis chat crippled your present life , and then you can do Tran somatic therapy that means that you talk to your wounds and come to terms that is better for the wounds and neurosis to go because as being discovered and is not the right time to be there to crippled your life, but generally the simple act of witnessing the wounds neurosis evaporated it dissolve it into cosmos is reabsorbed but they can be hard conditioning that need a dialogue then you can do Tran somatic therapy and dissolve them . Anyhow is a very enchanting psychedelic journey into cosmic unconscious specially to witness all of your past life in a retrospective journey backwards , first of all you get a revelation instant that death do not exist that is an evolutionary journey your consciousness, of refinement of your consciousness into light empty consciousness , and death is just a great fiction changing of house cloths changing of body but death do not exist , and this is a great revelation very important because when you don't fear anymore death in your life your life change into a peace blissful journey into eternity , no anguish no anxious anymore no frustration of die anymore you relax and the journey of your life will be a blissful journey into eternity and resurrection , And more is a very psychedelic journey you discover who you were in many of your past life and many event of those past life that maybe they crippled your present life and you clear up all of them through your witness consciousness , and your present life will be just a flowing of peace silent bliss sacred divine relax holy , and your unconscious just a pillar of light empty consciousness free absolutely free , And so you go on in the retrospective journey into the cosmic unconscious thousands of past life have happen , until the button , is very important that you remain with your witness consciousness awake present , then there is the animals layer when you were different forms of animals , the realm of animals hood , you have being many kind of animals tiger lion dog cat , etc.. that someone still give a certain in print blue print to your present life , and witness them you can record nice the animals that you were in your past life and this reflect even into your life today , anyhow then we get the nature layer the greenery plants layer you have being plants trees your consciousness into an unconscious stage as pass through the nature layer too and if ever happen that you do this inner journey at this stage is a rainbow vortex of colors spiral like ha is divine sacred psychedelic , then we get the rock layer were your consciousness at this stage is absolute unconscious asleep into the rocks but still present unconscious asleep , into the rocks consciousness is present unconscious asleep , in an evolutionary process towards refinement , then water consciousness is present into the water at a stage of unconscious but flowing through then probable the last layer of bacteria infinitesimal bacteria were the first molecule of DNA as take place 3 and half billions of years ago , and then the irrational jump into non being body incorporeal were time space forms and duality totally annihilate and you find your essence annihilate dissolve melt into super conscious , everywhere and nowhere in particular your essence are enclose into a circle of zero just like an infinitesimal atoms , at this stage is just bliss subtle ecstasy peace , no connection anymore with the body on earth if you still have one , and no will desire to come back into the body , is so blissful divine ecstatic you are in essence everywhere and nowhere in particular expand and intrinsic to whole existence into non being body just an I am ness a great relaxation ecstasy bliss , no will desire to come back into the body if you still have one , I have one young and beautiful but for sure I was not going to come back into it if casually a girl recall my essence into the body she touch my body recall me into the body and suddenly like a vortex my essence come back into the body with a tremendous rush of energy and I was completely naked into a forest , but what come back into the body was a new fresh presence of consciousness awareness fresh and every moment fresh since then and totally enlightened , I am talking by my own experience not by theory , Anyhow this is the journey in retrospective into the cosmic unconscious of the conscious body , don't emulate it or do it alone if you go for such a journey is important that if you are a man a woman is there with you that she can recall you back into the body or if you are a woman is important that a man is there with you that can recall you into the body otherwise you may never return into the body and left it forever out of great immense bliss , this is the experience of enlightenment express in detail , if your journey is done conscious with the witness consciousness awake when you return into the body enlightenment is yours forever , and you have burn out all of your past life and different layers of the cosmic unconscious , into the cosmic unconscious and into the collective unconscious and individual unconscious just a pillar of light empty consciousness is left and you are unconscious free forever and ever ....here end part 4 of the seven astral body of an organic unity follow next for higher body revelation ...Angelo Aulisa

The new book Mysticism & Physics Chapter 26 Angelo Aulisa

The seven astral body of an organic unity part 5

Hi friends , into the conscious body the size remain the same as physical , and basically is the quantum leap into the vertical dimension of deeper and higher the present split second is the gate less gate into your inner being , the conscious body is the dimension of being , you close your eyes bring the attention within yourself , and the gravitational energy field call dark matter will bend the light empty consciousness deep within your inner being , relax get centered rooted at the deepest point of your inner being , you are not your body not your emotion not your thought , not your unconscious staff not even your senses , but just a crystal clear witness consciousness that witness all and everything that surround you mirror like in the present the empty mirror reflect all and everything of the situation and next situation the mirror is empty again ready to reflect next situation , and at the deepest point of your inner being a pin drop silence is the gate less gate to your inner universal consciousness, and here the expansive energy field call dark energy will pull upwards your light empty consciousness up to the event of the universe on and on expanding it in synchronicity with the evolution expansion of the universe, and here you have enter the six body the cosmic body consciousness is the fundamental law of the universe the intrinsic fabric of life of the universe is just a pulsation of love intelligence peace a quality a creativity sacred holy divine , here on the overlapping of conscious body and cosmic body a break through happen majestic, Buddha call this breakthrough Anatta the annihilation of your inner being into the cosmic body , the size change into universal body size you in essence intrinsic one with consciousness are big huge as the universe is huge expanding every single second with it , because the universe is expanding every split second , and you flow expand with it and at this stage your inner being is the whole universal body , you are one with it in a mystical union intrinsic and indivisible intrinsic to all living being and forms of the universal body is a majestic divine sacred holy break through quantum leap mutation , and you flow expand with it and the flowing energy field called dark flux will flow move circulate your light consciousness in and out of the core and source of the mystery of the universe non being body incorporeal, in a festival of light empty consciousness I repeat at this stage your inner being is the whole universal body and the light empty consciousness move flow circulate in a cosmic play irrational in into your inner being that is the universal body, and out into non being incorporeal body and in this overlapping of universal body and non being body then nothingness the absence of things emptiness is the gate less gate into non being body incorporeal were time space forms duality annihilate completely totally, and also consciousness annihilate totally because consciousness exist always in relation to a subject or object into non being body no subject or object survive and consciousness annihilate into formless awareness relation less and unfocused awareness is the ultimate essence of non being body is just an I am ness a great relaxation, peace silence bliss into the law of eternity core and source of the mystery of the universe and of life and death and of all duality , eternity means no begin no end never born or die is the ultimate canvas of the universe huge bigger and vaster , beyond above transcendental to the universe itself, eternity is the ultimate canvas were the universe is display paint , and infinite open relativity an infinite freedom bliss peace an infinite silence intrinsic with subtle ecstasy that surpass all understand , the size into the law of eternity vanish into infinity , and here you are at home were the immortality resurrection reside take place unfold , the resurrection is a conscious alchemy from unconscious to consciousness to awareness into non being body incorporeal into the law of eternity not physical not gross not material, but just a conscious alchemy and here you can relax and the unconscious sleep and various hypnosis totally vanish gone, here you in essence are totally enlightened here enlightenment as happen forever and ever, you will be one intrinsic to eternity itself in a mystical union indivisible , here the ultimate aspiration mutation of your DNA unfold and take place immortality and your DNA as accomplish is journey of mutation and convergence of evolution in a triumph of civilization of the human being , and your consciousness totally refine into light empty consciousness as annihilate into light transparency, whiteness unbounded infinite light , and you can relax and have a cup of tea and enjoy enlightenment life become a cosmic play sacred holy divine a subtle ecstasy silence peace , here end this beautiful journey into the seven astral body of an organic unity man is not mere matter physical, is much more is an infinite journey endless into eternity without begin or end, actually it end nowhere ever man is a transcendence above beyond into the law of eternity core and source of the mystery of the universe and man is infinite significance juice playfulness a miracle in itself , to love to live it intense and with great zest ......Angelo Aulisa

Published on October 23, 2017 07:43 • 47 views

The new book Mysticism & Physics Chapter 27 Angelo Aulisa

Sabbath is made for people & not people for Sabbath

Hi friends , here I pay my respect to my beloved friends Jesus , he say many time that Saturday is made for the people not the people for Saturday , In the world today exist dozens maybe hundreds of constitution each nation as is own constitution all of them old and out of date expire bogus , a constitution is write down hundreds of years ago for the people who live that contemporary age , for them the people who live hundreds of years ago may have being good actual , but as time goes by dozens of years goes by 50 , 100 , of years goes by that constitution write down hundreds of years ago is not anymore actual the people have grown evolve in intelligence, much water as flow under the bridge, thousands of event situation happen all the time , physic science as evolve grow expand more in the last twenty years that in the previsions 3 hundred years, that is the age of physic science , or humanity as evolve grow in intelligence in human right technology more in the last hundreds of years then in the previsions ten thousand years , internet the web, we can travel the entire world in hours the simultaneous communication of media , the world has become a small global village , the DNA of human being as gone through hundreds of convergence of evolution mutation quantum leap just in the last hundreds of years then in the previsions ten thousand years , life is a constant continuity of change and changing is the only reality of life , there is only one thing in life that do not change and that id death , for everyone death is the only unchanging reality , death is the real communism the real communist were we are all the same , otherwise uniqueness is the truth, and until I will be alive I will assert my uniqueness everyone is unique in itself , the concept of communism that we are all the same like the assembly line of a car factory is a death concept . So life is a constant change mutation of convergence of evolution in intelligence in quality of living in shape of living , you should flow with the moment and consciousness moment by moment , Now this dozens hundreds of constitution around the world write down by the old ancient little men unconscious primitive barbarous insane sick , are expire out of date it may have being a good constitution for the men who live in the begin of the 20 century or in the middle of 20 century or 30 years ago 20 years ago but expire and out of date for the actual modern men who live our contemporary age 2018 , the politicians clever chaps they steak with a stereo type mind totally identify with the mind to the article of the constitution to exploit people to cheat people, this game as to finish , first of all constitution of the world dozens hundred are bogus false and out of date expire , a constitution is write by men in a particular frame of time in relation to the situation circumstance that exist in that particular time, and the men who write it down were unconscious primitive and with an ego mind unconscious abnormal , and a constitution is a document that should update every six months in the modern age that we live modify , is amazing how politicians clever chaps steak to the constitution to exploit people to cheat people , why should I live with a constitution write down by Mussolini , he was mad neurotic crazy psychopathic , Mass murder , a tyrannical dictator , for example I simple tear down that constitution and write a new one like Renzi was going to do , for example , and more all the constitution are expire and out of date because the modern actual men live a contemporary of age of 2018 were nothing is the same as 100, 50 , 40 , 30 , years ago all is change and the world is a global village and the UN and all the world should live with one world constitution one for all the world, an constant modification are included you can modify it up date it any time , because as being write by men and men can modify it any time , politicians clever chaps with the greatest inferiority complex of the world use the stereo type mind identify with that dirty paper this article and that article but if as being write by a man that by the way is death most probable , and is not actual anymore , as expire is out of date , it need up date to the actual situation , look the constitution is in service of men and not men in the service of the constitution if need to be modify change we can do in hours , the politicians clever chaps look at the constitution as if is a holy book with stereo type mind , but I tell you that the older is a constitution the more rotten out of date expire it is , the constitution as being write from a blind unconscious asleep man and can be and as to be up date every six months to be actual contemporary , same the holy books of religions the older they are the most rotten out of date expire full of lies superstition hypothesis hypothetical are , because they have being write by unconscious asleep primitive ignorant ancient barbaric people they are a bunch of neurosis a night mare of the past unconscious asleep ancient primitive men and you want me to live in 2018, in our contemporary modern age with a book write 5000 , 2500 , 2000 , 1300 years ago are you crazy insane sick or what, there is no trace of that primitive unconscious asleep men that have write such a books , and do you want me to denied all evolution in intelligence in science in physic in arts in literature in all the field of human intelligence in psychology , do you want me to denied all heavy sacrifice that human being have done to grow up in intelligence , and do you want me to denied all the convergence of evolution mutation quantum leap that our individual DNA as gone through to live a better life of quality, of good shape of contemporary to our age 2018 , to believe an old rotten expire out of date book of fancy tale for retarded crippled children full of superstition lies , hypothesis hypothetical , never ever I am a man of my contemporary age and I live contemporary to my age with an update consciousness awareness of my age , all holy books of religions are to be abolish ban stop are rotten full of garbage from the begin to the end they have being write when they was no television no radio no internet no web no science no physic no science of psychology and the geography of the world was not even known , and the universe was a sky of 300 stars that they were believing that they were lantern that illuminated the earth, and the were believing that the earth was the center of the universe , and was a flat planet , the earth is just a dot very small dot of the universe a grain of sand . Enough of this clever chaps politicians with the greatest inferiority complex of the earth that exploit people cheat people saying look at the article 155 but who care if they are 10 million people desire wishes that say is a laughing stock , the constitution is in service of the people not the people in service of the constitution and that document should be subject to update every six months if wants to be actual otherwise is a stench piece of paper old and rotten , But I repeat the world is a global village and there is a UN and they should be write a unique world constitution for the all world that I have given the guidelines to the UN to write down the new world constitution , they should write it down and all the world with functional govern should sign and subscribe the constitution , so then they cannot use their fancy constitution old and out of date expire to cheat and exploit people , with a stereo type mind look at this article look at that article but if they have being write by Mussolini who care about that fanatic dictator mad crazy man , are out of date expire , finally a unique world constitution with a strong unite nation where is imply all human right all right and freedom of all the people of the world were no gender no color no race no age make any difference, no discrimination of any kind , were the world is one global village and live through a single empty consciousness with no label no adjectives no contents but just consciousness fundamental law of the universe a pulsation of love intelligence compassion peace , just a quality creativity the very intrinsic fabric of life of the universe, that modern physic called Boson x the quality that hold keep the all universe together and confer Mass to particles atoms matter and without the Boson x the universe disintegrate instant , mysticism call the Boson consciousness the quality that confer organism to the universe oneness mystical union , otherwise we would have being the universe a sum of part mechanically assembled as the constitution around the world are mechanically unconscious assembled but is not so the world the universe is an enchanting organism a oneness a mystical union with the organic unity that human being are one indivisible , ...Angelo Aulisa

The new book Mysticism & Physics Chapter 28 Angelo Aulisa

Sabbath is made for the people & not the people for Sabbath part 2

Hi friends , so in short any constitution around the world are out of date expire and bogus , basically are made in way to exploit people to cheat people , because they have never ever being up date to the contemporary situation they were good for Mussolini Hitler Staling Mao Zedong , but fortunate humanity as evolve grow up in technology science physic intelligence man has come of age 2018 and there is no trace of the old serious unconscious ancient primitive men, men today live in a totally different dimension age time , the world is a small global village a dot in the infinite universe , it should be a unique world constitution one for the all world with a strong UN unite nation and all the functional nation who want seat at the UN have to sign and subscribe the new world constitution otherwise they can seat in the forest in the wild nature , because the game to consider the different constitution an holy book and to read with a stereo type mind totally identify, for cheat exploit people is finish end is not anymore possible we live in a contemporary of simultaneous global communication, were the fact are known simultaneous to the happen , and basically if you have never update that stench piece of paper call constitution is not actual anymore , even the new world constitution as to be made clear from the begin that updating are included any time, and necessary because the reality is a changing phenomenon every split second changing is the only reality , alive , and as the new world constitution as being write by men also men can modify update it any time is not an holy book but a document of sharing common sense of freedom meritocracy, I don't say democracy because democracy is expire out of date a default system that as fail all over the world a nick name for exploitation cheating people a system of corruption and dishonesty , so meritocracy freedom human right , and not discriminatory , towards genders color race ageing , and any time need to be up date modify is welcome the new proposal if is true and right , Is not a holy book that anyhow are all of them a bunch of neurosis garbage lies hypothesis hypocrisy superstition from the begin to the end dangerous for humanity because they make human being crippled conscious retarded hypnotize unconscious asleep , and they create conflict wars massacre bloodshed death destruction for futile motivation that exist nowhere if not in the neurosis of the little ancient unconscious insane sick serious men of the past , they are a bunch of past retarded tradition fancy tale lies , they are holy book that have being write by people who have commit genocide they butcher massacre kill all the previous culture population of Zaratustrians literally massacred all of them and on the ash of the destroy massacred culture they write the holy book . Basically all the constitution around the world all the holy books around the world are the power of argumentation , an argument is prostitute he it goes with the one who pay better , even if you are wrong and mistake with the power of sophistry argumentation you can't turn wrong into right for yourself, mistake into right thing for yourself , in the Ancient Greece they were school of sophistry were they teach people how to argue and debate any argument, and you would get the degree when you have the power to turn a wrong argument to the right for yourself and they were many master of sophistry that they learn that even if you are wrong is only a matter to pay better to discuss better the argument and you can turn anything to your own right, even if you were wrong an argument is a prostitute it goes with the one who pay a better price , today they are the lawyer that do this in regular base in courts , but sophistry in Greece was an early development of philosophy an early stage of philosophy, with the develop of philosophy sophistry was put out of date expire, they flower school of philosophy all over Greece because philosophy is irrational no matter who is wrong or right with philosophy you go beyond wrong and right and of course is an higher stand point , and people put out of date sophistry , But today days look like that sophistry is more powerful of philosophy because we live in a material world the dimension of having and horizontal , certain sophistry appeal more then philosophy , but one thing if you take an argument and take one thousand people and ask them one by one what is there interpretation they will be one thousand different interpretation each one will refer a different story, so certain that sophistry play the major part in the game who pay better the argument who argue better win the debate, but it do not mean that he is right most of the time who win the debate is because of cunning exploitation, is because he is more clever in cheating the people and he is a dishonest , so to me all constitution around the world are partial relative to the interpretation of part of who is spelling them so are the holy books . Anyhow is urgent a unique world constitution write down by the UN I have given already all guidelines in one of my book the world constitution the sooner will be write and be actual the sooner we will have a better world peaceful and free that live in meritocracy . without exploitation and cheating of people and peaceful civilize world for the first time civilization, the first principle is a world without weapons of any kind they are a crime against humanity no weapons no wars no military no bloodshed massacre no deaths killing, no destruction no starvation and misery, this actualization is a natural equation the outcome . More last happiness how to be happy in happiness , an happiness that as a cause is not happiness anytime the cause can be taking away and your happiness will turn into sadness, any happiness that as a cause is momentary to the cause is an effect of the cause , and can turn into sadness anytime , the real happiness is causeless is when you are centered into your inner being and your inner being is the source spring of your happiness nobody can take away your inner being , when you are centered into your inner being your true nature in essence you are pure love, the intrinsic nature of your inner being is love and happiness without any cause or purpose , you open your eyes in the morning your eyes meet the blue of the sky the greenery of the nature and the miracle of love happiness life unfold purposeless causeless, just because you are alive and full of zest , any happiness caused is temporary elusive momentary , and anyone can take away at any moment and you fall flat into sadness , the happiness the love that spring out of the source and core of your inner being is causeless is just the intrinsic nature of your inner being and nobody can take away and actually is called bliss, because as no opposite to it happiness as sadness as opposite complementary in a law of dialectic , bliss as no opposite complementary to it bliss is beyond above transcendental nature of your inner being as love too, you are love in essence you are bliss in essence, the secret is to be centered into your inner being and I have tell you in the previsions chapter hundreds of time how to be centered into your inner being , happiness is elusive momentary temporary in relation to a cause, bliss is causeless just an essence of your inner being it spring out of the core and source of your inner being and as no opposite complementary or cause , bliss is a synthesis in itself higher and beyond happiness and anyhow the real happiness is causeless an happiness that as a cause is elusive you take away the cause and it vanish into sadness ......Angelo Aulisa

Published on October 26, 2017 05:06 • 42 views

The new book Mysticism & Physics Chapter 29 Angelo Aulisa

Sabbath is made for people & not people for Sabbath part 3

Hi friends , so thesis sadness antithesis happiness friction of those synthesis Bliss, and bliss as no opposite complementary to it bliss is a stage of peace , love , silence with intrinsic subtle ecstasy , zest what means zest Marcelo ask zest means a stage full of enthusiasm for no cause at all for no purpose at all because zest is related to Zen to meditation zest is the stage of enthusiasm of meditation , compassion , harmony mystical union oneness with the organism of the universal body , balance in bliss you are either lopsided to the right or to the left but you are exact in the middle balance either to one extreme or to the other but just in the middle in equilibrium , relax calm , just a crystal clear witness consciousness mirror like an empty Belgium mirror , Belgium mirror are the best quality mirror of the world you were knowing it ,bliss is a stage of freedom from ego mind unconscious is a stage of no mind , bliss is attaraxia a stage were nothing disturb you were you are relax calm , balance in harmony in equilibrium Attaraxia is a Greek term meaning enlightenment , and bliss is the real happiness causeless purposeless , just because you are alive when you weak up in the morning your eyes meet the blue of the sky the greenery of nature the singing of bird ha this ha this and the miracle of life unfold moment by moment, an intrinsic miracle celebration of life every split second no past no future no present as you know it, but a present less present outstanding the 3 tense of time, the present that you know is a sandwich between past and future is condition by the past by your unconscious mind is dictate by your past and is a disease present and is condition by your imagination of the future, basic the past is dead is no more a dead tense and the future is not yet and your disease present stand in between as a slice of a bacon into a sandwich , the present less present that I mean is an outstanding present not related to the ego mind unconscious, which you are all loaded ,you can call it now ness or the moment flowing just now gone flows away, take a quantum leap an irrational jump into the flows river of consciousness awareness, and bliss is yours too , Heraclitus an ancient Greek mystic say that you cannot get wheat twice with the water of a flowing river, and I say into you not even once such is the flowing of consciousness awareness, which never stop not even for a split second to flow, consciousness creativity is a flowing quality pulsation that never stop to create not even for a split second , creativity evolution mutation quantum leap is the very core and source of the universe , there is no fix creator that create the world the universe in six days and then he went in vacancy and nobody as seen him anymore if it was like this it would be a dead static world , with no evolution of any kind , he would have create a perfect world universe, and perfection is death no more evolution , and take a look around yourself it look like a perfect world to you is worst chaos hell of imperfection , the universe the world the fundamental law of the universe the creativity the pulsation of love intelligence the intrinsic quality is empty consciousness and is a flowing like a river of evolution towards refinement of itself towards better quality of living better shape of living better intelligence is a constant flowing of evolution of consciousness, same as our DNA is a convergence of evolution mutation quantum leap of adapting coming in accordance with any situation circumstance, and towards a realization relaxation of itself the ultimate aspiration evolution mutation convergence of evolution of our DNA is immortality resurrection , all the seeking of men is not to know God but to be God itself , and our DNA is evolving and it will never stop evolving until the ultimate evolution mutation quantum leap as happen and the ultimate aspiration is immortality resurrection , Finally the mystery is what has no cause whatever as a cause is not a mystery we know the cause and the effect , the real mystery of mysticism is what is cause less and the core and source of the mystery of the universe and of life and death and of all duality the law of eternity that means no begin no end never born or die, is a transcendence beyond above the universe itself is the infinite ultimate canvas bigger and huger of the universe itself were the universe is display paint , as no purpose and not cause at all is cause less and remember what has no cause is the mystery , and it will remain a suffuse mystery for eternity to come , and that is the beauty because a mystery is unknowable and the unknowable make the life fresh every split second unknown nobody know what come next is a trilling every split second and intrigue of mystery , and that make life beautiful fresh in the moment flowing , otherwise it would be a mortal noia , boring , so the mystery is that which as no cause , and what as a cause is known , the mystery remain an unknowable mystery even if I have explain to you all the friends almost the all picture of the mystery but still keep a few cart hide in my harms , you know a good gambler he never show all is cart at once , and what I have given you is an hand full of dry lives but the carpet of dry lives of the all forest is still is laying with me , ha ha ha by hope that this great lesson of intelligence help you all to live a better life conscious aware intelligent ......Angelo Aulisa

Published on October 27, 2017 04:01 • 37 views

The new book Mysticism & Physics Chapter 30 Angelo Aulisa

Meditation & path

Hi friends , I would like to unfold what means to me meditation and the relation of meditation with different path , first of all what means religion , to put together to unite religion is a Latin word that means Religere to put together to unite in a oneness in a mystical union , Religere means what unite you in mystical union oneness with the organism of the universe , and soul means integration what integrate you into your inner being soul , your soul keep you integrate into one self , and once you are centered into your inner being you are rooted integrate intrinsic to your inner being soul , and your inner being is the bridge the link that integrate you unite you put together you in a oneness mystical union with the organism of the universe , your soul is the bridge the link with universal consciousness fundamental law of the universe intrinsic to the universe run a tread of pulsation of love intelligence peace silence a quality a creativity called empty consciousness, the very inner fabric of life of the universe that run intrinsic through all forms of the universe and living being sacred holy divine , modern physic call consciousness Boson x is the quality that keep hold the all universe together and confer Mass to particles atoms matter without the Boson x the whole universe and forms of it disintegrate instant , consciousness is the quality that confer organism oneness mystical union to the universe otherwise we and the universe would have being a sum of part mechanical assembled and the human being would have being a robot a machine , first you get integrate into yourself through your soul the soul cause your integration otherwise you would have being a fragmentation of thought emotion unconscious staff senses body , in fact your soul integrate you into an organic unity and from a stage of inner being you are a witness consciousness that witness body thought emotion unconscious staff senses at a certain distance that is why you can see them , once you are integrate rooted centered into your soul you are an organic unity one in mystical union with the organism of the universal body and an organism happen a mystical union a oneness happen , in fact I would call religion organism religion is an organism with the organic unity and the universal body , from the integration root of your soul you are put together unite one, in mystical union with the organism of the universal body and consciousness , it happen simultaneous once you are integrate into a soul simultaneous you are put together unite one in mystical union oneness with the all existence and universal body one indivisible , hence the word term individual means one indivisible with the all existence universal body , and an organism happen more correct an annihilation of your soul into the organism of the universal body and you are unite put together one in mystical union oneness sacred holy divine into an organism , that the meaning of religere , so soul means integration and religere means to put together to unite into an organism into the universal body in a mystical union sacred holy divine in bliss peace silence in a subtle ecstasy that the significance of religere , That why I am so against the common old out of date expire superstition hypothetical religions that exist around the world because instead of unite human being in a mystical union with the universal body they do exact the opposite they divide split human being finally and in an irreversible way, from the Cosmos which means harmony the word cosmos means harmony literally , they split human being from the core and source of l the mystery of the universe and of life and death and of all duality from non being the law of eternity , they make men anthropomorphic to believe in a personal God that do not exist, first of all is an hypothesis at the most an hypothesis , and they give image character and behavior to the person God , and this crippled make humanity infantile childish conscious spiritually retarded and most of all, they divide split human being from the core and source of the mystery of the universal body , they make human being rootless believing in fancy fairy tale , and this make human being rootless neurotic insane sick disease infantile childish , they do exact the opposite work of religere , because empty consciousness cannot have labels contents or adjectives they are partial interpretation of the little men, and as you give adjectives contents labels you create division conflict wars destruction death killing bloodshed , in fact this humanity as fight ten thousands wars most of them for ideological difference cultural difference of philosophy futile motivation and a conflict is on a war begin , The old out of date religions create chaos neurosis conflict wars, they divide split human being from the real true nature of himself and from the core and source of the mystery of the universe, all the old out of date religions should be abolish ban resign all together at once if we want a better world intelligent that evolve in peace in physic in science that evolve in freedom in spontaneity natural, sorry they are cancer for humanity a serious disease of neurosis enough they should resign all of them . That is the mean reason why I am against religion the split and division that they create, which is consequence of thousands of disease that exist today on planet earth. For example to me consciousness as nothing to do with violence weapons wars killing destruction , is the madness the neurosis state of human being that give an interpretation to religions as violence to kill in the name of a fictitious God that exist

nowhere absolute madness neurosis insane sick senile , is a partial interpretation of mad people that are totally identify with dead past tradition, because in there tradition the sword as being used to convert people with the treat or I kill you or you think as I think with terror, so they do in the present day with weapons this absolute insane violent crazy mad , but what to do their history is a history of wars terror killing butchering destruction since the early begin to the end of our day , To me this are not even worth of being associate to religion are mad psychopathic , religion as nothing to do with any kind of violence with any kind of weapons with any kind of wars or killing or destruction, interpretation with any kind of labels adjectives or contents , all interpretation are partial to the little old men out of date insane expire mad , sorry religion means simple RELIGERE to put together to unite in a mystical union oneness with the organism of the universal body and consciousness empty consciousness no contents no labels no adjectives and soul means integration to get rooted centered within one self , I admire Mahavira the greatest enlightened that as work on planet earth , for him even to breath was violence he use a rope in front of his mouth because is breath could kill some living being in the air and he was afraid to walk because he could do some violence by walking into some insect and killing living being , this sensitivity this organism this mystical union oneness with the organism of the universe and consciousness I call RELIGERE ,this nonviolence attitude is RELIGERE to be unite , one sacred holy divine this is my attitude too I feel tremendous affinity with Mahavira , for this I tell you, and for many other quality that he has he was very ascetic he eat once every twelve day and totally vegetarian he get nourishment from the air he breath and from love itself for example , this balance harmony , You know that the food today it carry all a percentage of poisoning contamination and to stuff oneself you accumulate layer and layer of poisoning that contaminate totally your body , all meat fish egg are loaded with poisoning of all kind , vegetable too they give pesticide , but in minor measure you wash carefully but they are contaminate too , to stuff the body is dangerous is like you accumulate multiple layers of poison , anyhow to be vegetarian and sensitive about food give you a better life believe it or not the lees you eat the better the less poison you accumulate . And last also all the martial art sword archery martial art to me are not related to meditation at all the ,subject of violence is in all of them I never get convince by the Japanese that martial art sword and archery can be meditation , can be concentration but concentration is not meditation concentration is the narrowing of the mind on an infinitesimal point or narrowing the mind into a Koan or object or subject but as nothing to do with meditation , meditation is a state of no mind no subject no object no Koan is the annihilation of the mind into the macro,into a no mid into a mindless consciousness or stage of inner being of witness consciousness crystal clear pure , into a stage of love bliss beyond above transcendental to any mind , so well know that concentration is not meditation and all this martial art archery sword ship first of all the subject of violence is intrinsic and is nothing else then concentration , if you talk to me not even Yoga is meditation the contortion of the body what meaning has to put your legs beyond the ears , yes they are particular position that you slow down the breath etc.. but to me all this afford is not meditation , meditation is a deep relaxation annihilation of the mind into the macro at the blink of the eyes no afford at all , and is pure love bliss ecstasy , yoga is concentration and terrible afford don't do the had standing it damage tissue of the brain is dangerous , is like taking cocaine that your blood circulate at hundreds of rate speed into your body and it flood your brain with immense flood of blood and it damage tissue nervous system and cell , same the head standing of yoga , to me yoga is not meditation it may be a good gymnastic but not meditation at all , finally a joke they was in the middle east in Samarkand a Sufi, that he had being in poverty all of his life then by fortune he get some money all at once , so the first think that he deed was to go to a tailor to order a new dress , the tailor take all measure carefully and he tell him come back next week and will be ready , the man happy waiting and waiting the next week he went to the tailor and ask for the dress , the tailor say is not ready yet I have allot of work to do but God will next week will be ready come back next week , the man went away and anxious all week wait then he went to the tailor and ask for is new dress the tailor tell him that unfortunate he as allot of work and the dress was not ready yet but God will next week will be ready the man up set anxious the next week come back secure that now is ready the new dress but unfortunate the tailor tell him the same story God will the next week will be ready, unwilling the man wait another week anxious and stress out he went to the tailor and ask for his new dress the tailor tell him unfortunate is not ready but God will next week will be ready at which the men fed up tell him tell me if we kill dispose of this God will how long it will take .... Same I tell you all if we dispose of God will how long it will take........
....Angelo Aulisa.....

The new book Mysticism & Physics Chapter 31 Angelo Aulisa

Meditation & path part two

Hi friends, so meditation is not concentration , narrowing the mind in an infinitesimal point or subject or object or Koan , that is a mind approach and mind is not meditation , concentration is a kind of self-hypnosis you exclude everything else and you focused on the object or subject or Koan that you chose , for example Mantra repeating endless a Mantra is self-hypnosis that induce sleep in the end repeating at nausea a Mantra you fall asleep, in a kind of trans that is self-hypnosis sleep induce because you get so tire of the repeating that same lulla by that you fall unconscious asleep , is not meditation that is full awake conscious beyond above transcendental to the mind into a no mind , with the blink of the eyes afford less you annihilate into the macro cosmos , into your inner being and simultaneous into universal consciousness, awareness into non being incorporeal body, no time no space no forms no duality of mind into the law of eternity, core and source of the mystery of the universe and of life and death and of all duality , an open relativity an infinite freedom, an infinite bliss sacred divine holy, endless begin less , never born or die ultimate canvas of the universe bigger and huger of the universe itself, were the universe is display paint into the law of eternity size vanish into infinity . repeating a mantra is pure self-hypnosis is narrowing the mind on those words you spell , and meditation is not controlling your thought or your emotion or your unconscious staff insight or your senses if you try to controlling them they will come rushing behind you with revenge , meditation is the realization that you are not your body not your thought not your emotion not your unconscious staff not your senses, through the centering into your inner being this realization happen, and you are just a crystal clear pure witness consciousness mirror like an empty mirror , an empty sky blue without clouds , the clouds are the thought emotion floating into your inner sky and if they are clouds you can watch witness at them, how long a clouds can stay into the sky a few hours a few days but finally they will pass and disappear vanish and the blue sky your inner empty sky blues shine again bright , you can only watch your tough, witness is the medicine fire that burn them off, through your witness watch at them they evaporate the cosmos reabsorb them , unfortunate in the west a wrong term is use for meditation because the word meditation imply thinking about, reflect about or contemplate about, meditation is not of any of these things mind activity , and the word meditation unfortunate induce to this conclusion , the right word for meditation is Dayan that in China become Chan and in Japan become Zen and Zen means literally meditation but it do not induce you to think other interpretation is simple an irrational terms words that come from the Sanskrit word Dayan, and every word term in Sanskrit as twelve meaning but none of the meaning is thinking reflecting contemplating is simple an irrational word Zen that means no mind annihilating the mind into the macro cosmos, a state of bliss sacred divine holy , that as no relation with violence weapons wars killing destruction bloodshed massacred far out , Zen means annihilating the mind into the macro cosmos and cosmos means harmony literally the universal body , so Zen means annihilating into Harmony the cosmos , is a love act is a love affair with the cosmos peaceful holy sacred divine, a bliss a subtle ecstasy a pin drop silence that surpass all understand . But unfortunate the world as gone totally insane neurotic mad crazy , they have make a mess confusion and everything is upside down , the other way around , finally Zen is peace love bliss compassion harmony, subtle ecstasy pin drop silence holy sacred divine and as nothing to do with violence , wars weapons killing destruction bloodshed massacred , far out come back to your senses ....And unfortunate the world is fragmented in hundreds of piece like a broken mirror and each piece of the mirror claims that is the truth , bullshit lies and lies and neurosis , the empty mirror whole and some is the truth, no interpretation of the empty consciousness awareness are possible they fall all short partial , no labels no adjectives no contents , of the little men , insane sick neurotic . Consciousness is simple a fundamental law of the universe intrinsic to the universe, a pulsation of love intelligence sacred holy divine, a quality a creativity that run through all forms and living being of the universal body the very inner fabric of life of the universe , that physic call it Boson x the quality that keep hold together the universe and confer Mass to particles atoms matter , and without the Boson x the universe disintegrate instant , consciousness the Boson x is the pulsation quality that confer organism to the universe, otherwise the universe would have being a sum of part mechanically assembled and the organic unity that man is a machine robot . No interpretation of part are possible no labels no adjectives no contents are possible is to infinite to enclose in stupid philosophy or religions culture, it diminish to define it with a partial mind infantile childish superstition anthropomorphic hypothetical hypocrisy. The world is just a small dot into the infinite universe and consciousness is expand everywhere and nowhere in particular into the whole universe millions of galaxies stars planets is infinite infinity, and into other galaxies planets for sure exist other organic unity and they simple call consciousness empty consciousness they don't know anything about your Jesus Christianity Hinduism Buddhism Muslim etc.. when you will meet them the aliens they will laugh at you they will tell you we never hear of Jesus , but we Know the Boson x consciousness the fundamental law of the universe that we know, and into consciousness we are all one mystical union oneness on consciousness we are link we understand we are bridge with you inhabitant of planet earth , but unfortunate we know that you are primitive barbarous unconscious asleep divide split in hundreds of fragmentation of superstition hypothesis infantile childish neurotic insane sick ideology, that you call religions , and for this you are dangerous we are not interest in relating with you inhabitant of planet earth you are neurotic insane mad crazy anthropomorphic and dangerous , sorry we are not interest in relating with you , maybe after hundreds of years when you will grow up in consciousness awareness we will be interest , that what the aliens will tell you . I say this because into this world today there is a hell of conflict wars violence madness neurosis between all nations and different neurotic culture religions and I include Das Capital of Marx as a religion the worst kind of conditioning exist on earth today , you fight make wars conflict for small piece of land for conquering this nation or that nation small piece of land stupid sorry idiot , when the world itself is just a small dot into the universe , planet earth the world should relate itself with the universe in relation of the universal body also planet earth exist , in an organism with the rest of the universal body where for sure is just mere mathematics other planets with organic unity exist , and give way to the exploration of the universe , give all resource available , Look today days as we are were humanity as reach into the universe we have explore just a little small piece of the universe very small points of the universe is known partially even , and unfortunate in the infinity immensity of the universe as we are today is impossible that we will be able to explore the universe at large , they are no space craft no vehicle and it will never being space craft or vehicle , because the distance are so huge years light away that even if it will be space craft vehicle we will die before to reach the destination unfortunate . We human being we know almost nothing of the infinity of the universal body nothing and the saddest thing is that probable we will never know. But they are some good news , starting from the entanglement of two particles locate far away from each other in different galaxies that they are simultaneous in communication communion and mystical union oneness what is the tread that unite them the mysterious tread that unite them , I have said in the previsions chapter consciousness make them one unite indivisible in oneness in a mystical union simultaneous , called consciousness the Boson x , so going deep into the mysterious tread that unite them and going deep into the cosmic web that exist into the cosmos there is a spider web into the cosmos made of dark matter gravitational energy field dark energy expansive energy field and dark flux flowing energy field , the definition of this 3 law of the universe will bring the greatest revolution into our planet and universe , that what all physic and science are working hard today to do to define this intrinsic mysterious law of the universe , so going deep into the definition of the entanglement of two particles distant in different galaxies from each other that means going deep into the definition of the Boson x consciousness , and going deep into the definition of this 3 law of the universe that I mansion , and defining the cosmic web more accurate , can open the way for human being to explore the entire universal body , through tele transportation , we can have a passport of our molecules structure and our DNA structure and move our body were we want through give the coordinate were we want go , using the cosmic web well define , the definition of dark matter dark energy dark flux and the Boson x consciousness is the definition of the cosmic web the spider web of the cosmos , and we can use as path to move into the universe is little beat futuristic but is the only way for human being to move through the space of universal body in real time simultaneous , but there is a danger that the definition of dark matter and dark energy give into the hand of this mad neurotic humanity of today of crazy politicians crazy religions that they will make weapons out of the definition and this will be very dangerous because intrinsic to dark energy and dark matter there is a potential of energy infinite immense that the nuclear energy of today will be out date as a laughing stock this is the dangerous maybe the world is not ready to receive the definition secret of this to mysterious law of the universe humanity is to unconscious barbarous primitive, they will make weapons out this definition , so is better that the definition is late until this humanity will be ready to receive the definition of the greatest secret of the universe , we need a humanity conscious aware peaceful were weapons wars military nation religions are totally abolish ban , and the earth is living a new dawn of consciousness awareness civilization in peace in intelligence, with I repeat no weapons no wars no military no nation no religions then the secret of definition can be deliver safe and will change completely the face of planet earth and the universe it will be the greatest revolution quantum leap convergence of evolution that humanity as ever witness but we need this earth the lotus paradise and this body the very consciousness awareness to deliver the definition , we can explore travel the entire universe in real simultaneous time , we can chose the galaxies that we want live I will chose sombrero is so mysterious and beautiful ha ha ha it will happen it may be hundreds of years but will happen .....Angelo Aulisa

The new book Mysticism & Physics Chapter 32 Angelo Aulisa

Meditation & path part 3

Hi friends , then meditation Zen is the bridge the link that create a mystical union a oneness of the organic unity and the organism of the universal body cosmos , whit Zen the individual the organic unity annihilate is own inner being into the cosmos into harmony and consciousness then an alchemy of oneness organism happen , an individual that means one indivisible with the cosmos universal body and consciousness, merge and melt into the organism of the universal body intrinsic to universal consciousness, then non being incorporeal body no time no space no forms no duality of mind , then formless awareness relation less unfocused, ultimate essence of the law of eternity begin less endless never born or die, an open relativity infinite , an infinite freedom an infinite bliss peace, and a subtle ecstasy intrinsic to pin drop silence that surpass all understand , holy sacred divine , into the law of eternity the size vanish into infinity, the ultimate canvas of eternity is bigger huger then the universal body itself eternity is the ultimate canvas were the universe is display paint , and is an eternal journey that end nowhere endless , into resurrection immortality , but still on and on intrinsic to eternity itself and intrinsic to the universal body intrinsic to consciousness as essence fragrance quality data intelligence you flow eternally for eternity to come , the resurrection is a conscious alchemy from unconscious to consciousness awareness, into non being incorporeal body no time no space no forms no duality into the law of eternity core and source of the mystery of the universe and of life and death and of all duality . As in Zen an organic unity and individual is universal relate to the universal body and consciousness, so I see planet earth in relation to the universal body because really this beautiful planet green called earth is a very small dot into the infinite infinity of the universal body were millions of galaxies planets stars exist into an enchanting organism , and planet earth exist into this enchanting organism called universal body , is an infinitesimal small dot , and that is why I see the world as global world a small village , today the evolution in technology the web the simultaneous communication of media the travelling in hours of the all world the evolution in intelligence and the evolution of our DNA as make of the world a global world a global village , and the way forward is a strong unite nation UN a unique world constitution , and a new dawn of civilization intelligence Zen meditation consciousness awareness, for the entire world a unique universal consciousness fundamental law of the universe that run intrinsic to all forms and living being of the universal body, and beyond the forms we are a oneness an oceanic consciousness a mystical union extension of one another sacred holy divine , a new dawn of civilization consciousness awareness intelligence that man will face alone, the sunrise of consciousness, man into consciousness is a light into himself , we don't need any more institution called out of date expire religions man has come of age and intelligence he will face alone the new sunrise dawn of consciousness , Zen the path or silence or love or painting or singing or dancing or sculpting or any creative path of harmony love that bring him beyond mind into no mind into consciousness , the path of Zen is the path of harmony of love of merging melting annihilating into the cosmos into harmony , is a path of love harmony flowing intrinsic to consciousness moment by moment into the present is not a path of power, Zen the word power and Zen are a contradiction in terms , Zen is a path of mystical union organism oneness harmony love with the organism of the universal body , no violence no weapons no wars no military no nations no religions no family no hard institution that bound freedom and spontaneity and naturalness will survive in the future, all the prisons will be shut down substitute with academy of meditation , were the person will be give the alternative opportunity to become an individual conscious aware, in harmony with the universal body and consciousness , and the family will be shut down in the future substitute with love inner love, the love of your true nature of your inner being in essence you are love within your inner being, this love will substitute the family which is an hard institution that deprive of freedom and universality the individual , no hard institution will survive in the future, in a new dawn of civilization intelligence and consciousness awareness , I see planet earth in relation to the universal body and as a global world , and the fight wars conflict that are going on for futile motivation of difference of sick insane ideology religions, are neurosis madness insane , and the conflict wars that are going on for conquering this piece of land or that nation are absolutely idiotic stupid the world itself is such a small dot into the universe , unless you want denied all the evolution of intelligence science physic , and all the evolution of mysticism the science of the inner mystery of man , unless you want denied thousands of years of evolution of our DNA that as had thousands of convergence of evolution mutation quantum leap , in better quality of living better shape of living better intelligence , unless you want denied all literature , unless you want denied the reality that our beautiful green planet is a small dot interacting into an infinite infinity called universal body , you know a small variation of orbit little beat far away from the sun and our planet will become a snow ball a little variation of orbit closer to the sun and our planet will burn out and all life vanish our planet is a small dot interacting with a huge immense infinity called universal body , a little variation of atmosphere and we cannot breath anymore and life will disappear or a meteorite falling on earth and will pill earth like an orange , I would like that humanity understand the inter relation interacting of our planet into the universe, and that humanity consider our planet in relation to the universal body , all the division split of ideology culture religions nations are over our intelligence as grown so much that this kind of conflict between nations for conquering of land power or wars of religions for difference of ideology culture philosophy are idiotic stupid childish infantile out of date expire , we live into an organism called universal body infinite you cannot measure infinite infinity and our search is for the in measurable the infinite infinity for the causeless what has no cause create the mystery what is in measurable create the mystery infinite infinity ......Angelo Aulisa

The new book Mysticism & Physics Chapter 33 Angelo Aulisa

Meditation & path part 4

hi friends , now I would like to unfold my understand and vision of prayer, the common gibberish talking non sense means gibberish that all over the world people are doing every day , if you talk with psychologist they will tell you that prayer is talking to yourself or at the most to the empty sky , is a form of disease as God do not exist is an hypothesis at the most an hypothesis a superstition that the tradition of the past goes on passing it to the next generation call anthropomorphism ,you personalize God , very dangerous disease because it create wars disaster and deaths killing , God is a wish fulfillment of the little old men need a father figure in the sky until he is a child he as a father then the father die, and the old out of date religions provide a father figure in the sky a mighty father , childish infantile , the believe in God is anthropomorphism a disease that give image character behavior personality to God they give personality to a faceless reality that is impossible , a disease that make men infantile childish conscious retarded and crippled for the rest of his life , with inferiority complex guilty complex , superstition prejudice unconscious asleep hypnotize, an irreversible disease worst then cancer, in my understand, prayer is a disease you talk to yourself to the sky as do not exist any God , an hypnosis of the old out of date expire religions go on perpetuating unconscious asleep to humanity, damaging hurting humanity making humanity childish, infantile, superstition, prejudice, crippled enslave to old out of date fanatic tradition , conscious spiritually retarded , with inferiority complex and guilty complex an irreversible cancer , the mighty God seating in a golden throne in the sky with white long bird this is the idea that most of humanity as of God, don't laugh is dramatic , and you small puppet on earth praying talking to yourself to the sky is dramatic . I will tell you a story once and one time in the ancient ancient past, God use to live on planet earth outside , and you can image millions of people in line crowd every day the court of God , and everybody everyday was asking God favors please give to me a good job , please make me wealthy please give to me a beautiful son , please find a job for my son , please find a good husband or wife for my sons , please make the rain come into my field , etc.... and every day for years and years millions of people asking and asking that what prayer is asking God for favors asking God your wishes desire as if God do not know himself what is good for you , and so God get exhausted tire stress out , and finally he call a meeting of is wiser men of the court , and ask this question to them I am so tire exhausted and I would like to escape some were , were should I go , immediate one wise man of the court say you can hide on Everest but God say but I know that sooner or later a man will come also there call Hilary , and then another man say you can go on the moon you are God you can do that, but God refuse the idea of leaving is own good people is own creation , and then the wisest man of the court stood up and say dear God you can hide within man itself there nobody has ever look, and nobody will ever look , and it say that God enchanting of the idea disappear and annihilate hide himself within men itself , and today is laughing at all the people prayer that they go Churches temple ask for this and ask for that and he is hide within them self-laughing , that what common prayer is asking God for favors asking God to do this and that , as if the mighty God don't know better than you what is good for you , and of course nobody is looking within himself that what meditation Zen his, and the few people who have dare to look within all have found him the mighty God because God is hide enshrine within your inner being soul , now this is a tale there is no God exist no God but if you look within yourself into your inner being love is find as the very essence of your inner being, as the very true nature of your inner being, you in essence are love and love is an inner quality of your inner being, and love is God , and if you look within your inner being bliss is find and bliss is a stage of happiness without opposite, bliss is God , and peace is find and silence with intrinsic subtle ecstasy is find within your inner being and silence is God and compassion is find and compassion is God and harmony is find and harmony is God and balance is find and to be in the middle not lopsided to the extreme Upeksha indifferent to the extreme in equilibrium is God , just Look within your inner being and your inner consciousness and all the treasure of life are there waiting for you, precious stone gold diamond platinum look within, the treasure is there waiting for you since ever, is a good deal you drop a heap of garbage your ego mind unconscious and you earn the all existence, the all universal body the all eternity the immortality the resurrection , you lose nothing and you earn whole , Zen meditation is the path of now and of the future , prayer is insane sick a pathology a disease a neurosis a night mare it do not exist any God you are talking to yourself to the sky, you are mad crazy ask the psychologist they will agree with me ....Angelo Aulisa

Published on November 06, 2017 00:55 • 54 views

The new book Mysticism & Physics Chapter 34 Angelo Aulisa

The generation gap

Hi friends , after a long break I would like to spell again the generation gap , this chapter you will find if you are interest though all my books I have write 14 books and I have write very accurate the generation gap chapter already many time , but seems and it is the fact that men never learn , men never learn scientist and psychologist say that only two per cent of what as being say the mind allowed assimilate to enter into the mind , true and even if enter into the mind within short is totally forget , and this is good in my own understand is a safety measure to keep non sense garbage out of the mind , basically everybody is doing mono log is talking to himself nobody is hearing anything , and this is good I repeat is a safety measure only two per cent the mind allowed in , basically everybody is doing gibberish ,talking non sense unknown language throwing into the air the folly, the unconscious staff of is ego mind , and the other is just a device to downloaded all the folly absurdity of your unconscious ego mind . Once they was a psychologist in a psychiatric hospital , and was watching two crazy men talking to each other , and he was surprise because when one was talking the other was silent , and then start the other talking and the other one was silent , so he become very curious and he went close by to hear what they were saying, and then he realize that the talking was not related to the conversation each one of them was talking something that was not related to the conversation , so he could not resist and he stop them and ask sorry why wen one of you talk the other is silent if the talking is not related to conversation, the two crazy men start to laugh and they answer do you think that we don't know the regulation of conversation , when one is talking the other as to be silent , but their discussion was not related each one of them utter simple gibberish , and that what happen all over the world , people talking to them self in a mono log , no dialog as ever happen , and that why men never learn don't think that I am exaggerating is the simple truth , no one is listen at the most he hear that is quite different , hearing is from the mind ego unconscious , listening is from your own being beyond mind ego unconscious and is totally different , that is why many time I am reluctant to write post or chapter, is useless wastage of time energy men never learn and nobody is listening anything never , and the unconscious ego mind prevent any listening it knows already , for example you can tell to a beautiful lady let go and see a sunset and she will tell you I have already seen the sunset , no matter that each sunset is unique in itself the mind ego unconscious knows already everything and it prevent any new happening any new exploration any new event any new life experience to happen , the ego mind unconscious is really the biggest disease that as happen to men , is a death before that you really die you already death ,because it cause the death of the cell of learning into the DNA, it prevent any learning any listening any new event or experience or exploration . So many time I ask myself for what go on writing is absolutely useless , like shouting in a desert , but I just see on my blog post that I have an average of 35 people views on my blog post I am surprise so I will write again the generation gap a very important delicate subject, it will take many chapter so get ready is gone be shocking for many of you , but if have the patient the love for wisdom and go through the next chapter, following you will got born again promise , I will enjoy delight in writing it down, but for sure many of you will feel hurt very much, so is not going be easy for you but for me just a pleasure, so here follow the generation gap , please listen please learn the next chapters follows are going to change completely your life, in the end you will feel thankful to me but only in the end mean while I write you will get very upset hungry frustrate in an anguish, but what can I do truth hurt, many time is a big pain but worth to go through because you will grow up, instead of growing old as you are doing you are only growing old and die and never grow up, is very sad is time to grow up higher and higher in consciousness awareness here follow ......Angelo Aulisa

The new book Mysticism & Physics Chapter 35 Angelo Aulisa

The generation gap part two

Hi friends , on BBC they pass one promotion were they say this is our history , I am very sorry but this is your history not our history , the spot show dramatic recent history bloodshed and massacred , I totally split tell apart from that history I am not part of it, basically I have done nothing to create such an history , those people religion that make that history are crazy fanatic psycho neurotic , totally mad , blind believer in tradition out of date, and religion out of date expire , were every 15 days start a new train of war conflict killing butchering massacred bloodshed , for them life as no value , are crazy mad , neurotic , that means that their neurons have being feed with wrong information notion, of life their neurons are perverted full of garbage of a philosophy culture that is the culture of wars killing bloodshed , and you know well that neurons function as vector of believe emotion thoughts in a chain reaction into the brain , and their brain is neurotic crazy psycho, disease with that wrong information notion philosophy of life , that is your history not my not our , that is your history of sadomasochism were life is value zero zero zero , were killing is the norm were wars are just the norm, since the begin of that religion to the end an history of war killing bloodshed destruction butchering of people a disaster , I totally stand away far away from that history, and is not my history at all , My history if you want use this word term, is an history of meditation, music, dance, painting, color , celebration , of peace of silence , of love , of delight into the miracle of life, sacred divine and life is sacred divine every living being is sacred divine holy , even insect , is a living being sacred divine holy , my history is an history of meditation of being in mystical union with the universal body and universal consciousness, and in mystical union with the core and source of the mystery of the universe and of life and death and of all duality which is eternity itself , which means no begin no end never born or die infinity , the size of eternity the canvas of eternity vanish into infinity , and me we are in essence one intrinsic one with eternity, the law of eternity immortal resurrected into the body yet , enlightened into the body yet , my history our history of the world of Sanyasy is an history of rejoice, delight in the miracle of life , of love the miracle of life , of music, dance, playing music, dance as meditation , of singing , of painting sculpting , of love each other as a sacred play of life , and life itself is sacred divine , into the forms we are seven billions but into the within into consciousness, we are one oceanic continent of consciousness, oneness, extension of one another , sacred holy divine , hurt an extension of your consciousness and you have hurt yourself indirect, kill an extension of your consciousness and you have commit a crime against humanity , our history of Sanyasy is a totally different plane level of living with a totally different focused gestalt , we love life , we love human being we love the planet, we love the universe, we love the core and source of the universe eternity itself , our history of Sanyasy is not a history is just purana essence, of here now in the moment living the present, moment by moment, flowing in consciousness awareness from a plane level of inner being , not as the crazy mad psycho monsters that they live from a plane of ego mind unconscious , totally identify with the ego mind unconscious, tradition, out of date religions out of date expire that lead this people to conflict wars massacre bloodshed killing, butchering starvation misery, that is your history not my not ours sorry please up date that promotion spot , we me we may be not so many maybe ten millions into the world, we Sanyasy we have no history but just Purana essence, of love celebration peace joy delight rejoice into the miracle of life, we called it love , and meditation with cap M is our tread bridge to be in mystical union with our inner being, consciousness , non being incorporeal body ,with awareness with eternity itself , our Purana is of dance, music singing , peace love compassion, of silence that is the universal language, silence is the universal language, a zero can merge into a zero and silence can merge into the universal silence AUM soundless sound of the universe, and be one with it intrinsic one in essence a human being is eternity itself, immortal resurrected, enlightened into the body yet , our Purana is an eternal journey into eternity without begin or end , our Purana is an everyday blessing which is beyond happiness , you know happiness as an opposite into sadness any moment can become sadness , our blessing is without opposite eternal, we live in infinite freedom real freedom, beyond the body, beyond the earth, beyond the universe, into the infinite canvas of eternity which is the ultimate canvas of the mystery of the universe, huge bigger than the universe itself, basically eternity is the canvas were the universe is display we are beyond into the core and source of the law of eternity, immortal resurrected, holy sacred divine, into infinite bless , immortal in essence , life is an intrigue mystery of bless celebration love, music color dance, playfulness delight rejoice, a joy to be present to the present, and always sacred holy divine in a bless , please to the BBC take back your history is not ours my never

ever never ever, I repeat if you are crazy neurotic disease mad is not my responsibility you take it back is not our history never ever ....ok Angelo Aulisa

Published on December 04, 2017 22:32 • 41 views

The new book Mysticism & Physics Chapter 36 Angelo Aulisa

The generation gap part 4

Hi friends , first a joke , demilitarize zone North Korea they was the captain and two soldier up to the wall , the captain tell to one of the military to go down the wall and check the fence down because it look loose , one of the military went down and check one of the fence , and he shout all ok at the first fence and shout do you see me now to the captain , the captain shout back yes , then he went to the second fence and shout all ok at the second fence do you see me now to the captain , and the answer was no , and the military answer back great and you will never see me again running in is way to freedom ,,,,,,Friends until the society around the world remain as they are , totally unconscious , the blackmail and rape of woman will continue increase , we need to abolish religions all of them around the world we need to change society asking the resignation of all politicians they are all corrupt and active doer of abuse and blackmail of woman , we need to abolish the family culprit of the disease call abuse rape of woman , and we need to abolish nations all around the world a strong UN is enough , and all functional nations related to the strong UN , and we need to abolish all weapons deadly and conventional absolutely and we need to abolish all military around the world , a new era of peace love honesty civilization , intelligence , meditation , inner being consciousness, non being awareness into the law of eternity, one consciousness for all universal body neutral to Gender color race age , no label no adjective no contents , the truth as it is no more lies , authentic , Is useless that you get shock every time you discover a new monster that rape abuse woman , he is a victim of your unconscious neurotic society , the sexual energy of a human being is like a river rushing to the oceans and if no hindrance are made it will reach the ocean of consciousness without perversion , but the religions the society the politicians the family are creating all sort of conditioning and freedom spontaneity naturalness is destroy , No dam as to be built on the river flowing towards the ocean ,who make the dam is responsibility of the disease call rapes abuses, of the neurosis that humanity is going through enough of religions, system society , family , nations , military wars ,weapons, bloodshed deaths destruction , we need to change 180 degree and make a new dawn of meditation, consciousness civilization, awareness non being into the law of eternity, to happen, intelligence compassion love peace, to happen.....Angelo Aulisa

Published on January 14, 2018 23:53 • 29 views

The new book Mysticism & Physics Chapter 37 Angelo Aulisa

The generation gap part 3

Hi friends , I am sorry but I herald a new era of peace , love understand , I herald a new dawn of meditation civilization inner being , consciousness , non being , awareness into the law of eternity , I herald intelligence understand love , dance , playfulness , so I repeat , abolish ban of all religions around the world out of date expire they preach only lies and they make human being anthropomorphic, a disease that give character behavior image to God, that personalize God , they create a face to a faceless reality , as consciousness is , consciousness is just a fundamental law of the universal body intrinsic to it , is just a pulsation of love intelligence is just a quality a creativity intrinsic to the universe , no labels no adjectives no contents can define consciousness , is the very inner fabric of life of the universe , physic call consciousness Boson x a quality that keep together the whole universe and confer Mass to particles atoms matter , without the Boson the whole universe and form of it disintegrate instant, is what the ancient little men call God , just update , mysticism as call the Boson consciousness since thousands of years , The religions of the world are hiding this fact reality because if it come to the surface the truth they have to give resignation all of them, and what they don't know is that all the intelligence of the world is waiting the resignation of all fanatic superstition out of date expire religions , when you feel like your resignation of all religions are welcome . and I herald the abolish of nations world wide, the world has become a global village small a dot into the universe , they will be functional nations but all related to a strong UN, with a new world constitution that all the functional nations have to sign and subscribe , if they want seat at the new UN , I herald the resignation of all politicians around the world they are all corrupt and active player in the blackmail and abuse of woman, and graveyard digger due the selling and dealing of the weapons all around, the world they are broker of weapons and they have dig a big graveyard for the all world due the sale and dealing of death device like nuclear weapons, and conventional weapons and chemical weapons , is a crime against humanity , to hurt or kill an extension of consciousness one for the all universal body is a crime, this people politicians have to be brought in front of criminal court for uncountable crime against humanity , the most dangerous men in the world today and dictator tyranny are Putin , Assad , Erdogane. And Trump he sale and produce weapons all over the world , death device he want produce more nuclear missile for America , this have to be brought in criminal court and charge as crime against humanity , and of course they are many other criminal leader dictator tyranny all over the world one dollar a dozen , we kindly ask your resignation all over the world , because the world would like to evolve to go on in intelligence in consciousness awareness , you please get update with the new world constitution , to produce deadly device call weapons is not allowed anymore , same dealing with death device sale and buying is not allowed anymore is a crime against humanity , humanity as evolve grow in intelligence consciousness civilization as had quantum leap convergence of evolution of our DNA and this activity that you are doing are not allowed anymore . Same the religions culprit of the neurosis of this humanity, celibacy is a crime against humanity , I wonder how this humanity is tolerate all the crime that religions are perpetuate since century , they abuse children the priest, they make of the human being beast that abuse rape woman, with their preaching of lies celibacy , they have the responsibility to have create a humanity of unconscious repressed conditioning human being , and they treat woman as inferior being, without a soul woman are not spiritual for religions , ashamed on you , I wonder humanity is ready to tolerate and accept all this misery lie bullshit, but is not ready to change towards truth that many enlightened one have shown up , is an addiction to misery and suffering , weak up before that is to late all religions have done there time is over give way to physic science and mysticism, meditation consciousness awareness, non being into the law of eternity , they are way path of living true authentic and real , Same the family is the most repressive conditioning institution, that create prostitution sex slavery , abuse of woman and rape of woman, the family is over as done is time too , I see a humanity of individuals centered into is own inner being and consciousness, living in love, and sharing their wisdom with the beloved and happy to be all alone all one with the universal body, with universal consciousness, and annihilated and one into non being body incorporeal body and awareness, and basic one intrinsic into eternity begin less endless never born or die , immortal resurrected . The family is a small unity that crippled the new man in is freedom, spontaneity, naturalness into is universality , the future is gone be millions of community, Zen field academy all around the world , and into the community man rediscover is universality and is own inner being of course, and live through it , your inner being is the bridge the link from the known to the unknown to the unknowable , is the bridge link to consciousness, non being, awareness, eternity itself core and source of the mystery of the universe and of life and death ad of all duality , the future belong to the individuals which means indivisible one with the core and source of the universe eternity itself, which is the ultimate canvas were the universe is display , and eternity is huge bigger vast of the universe itself, is the ultimate canvas no forms no time no space into eternity, the size of eternity vanish into infinity no size no begin no end . I herald the abolish of military and weapons, factory producer and dealers of such deadly device they are making crime against humanity, is not allowed anymore in a new world constitution, we the world kindly ask there resignation now finish, you are making a crime against humanity you can find other items to produce start from equipment to reduce pollution all over the world, why produce items that kills other human being extension of your own consciousness is a crime. You would say but this is a Utopia but I am an incurable dreamer, of this planet the lotus paradise and this body the very consciousness awareness. Either you got this point now or it will be to late, we are running out of time the Eco system is collapsing the world is active with dozen of wars massacred bloodshed death destruction , enough , please update your consciousness awareness and make some real steps towards this transformation, towards this quantum leap convergence of evolution , the woman they should stand in first row , are you tire of being abuse rape, treat as you have not even a soul, so assert your human right now , and be the herald of this changing of a new dawn of consciousness awareness, civilization intelligence compassion , is now or never urgent we are running out of time, and we cannot tolerate such a state of affair any longer, right you all woman, I guess you agree by love all of you get courage ......Angelo Aulisa

The new book Mysticism & Physics Chapter 38 Angelo Aulisa

The generation gap part 5

Hi friends , the title the generation gap part 5 but I have not yet begin to talk about the generation gap ......soon I will maybe 19 ,01 2018 ,,,, the resurrection ...the Samadhi ,,,of the emperor of mysticism Osho , never born never die no begin no end he just visited planet earth a divine sacred holy passage , I we love U Osho forever and eternity to come until my last breath will be love Osho , your dream is alive more than ever we will make it real true 100 per cent soon , the old men is on the death bed , they are few megalomaniac still that they think that the world is there private property, fortunate there is death soon they will be gone a split second and they are gone , and the world will be free of the worst ignorance superstition ever witness , actually the dusk the dawn is close by but as you know there is a moment before dawn that is so dark that it look like that the dawn will never come , but finally the dusk the twin light and the dawn the sun on the horizon start to rise , I have live thousands of night like this waiting for the dawn the dusk and really there is a moment when one last hope and is so dark that u start think that the dusk the sunrise is not going to come, but then the twin light and the dusk rise up the liberation, and the beauty the color the fresh breeze the ocean, the palm trees the music divine, all the beloved dancing at the best, a universal orgasm of joy bliss ecstasy ha this thousands of time ha this , that is it , the dark night of the soul is at this stage , near close by to the sunrise , but still they are few megalomaniac that hinder the process , dictators tyranny , that they think that the nation they belong is their own private property , idiot sorry, you are there for a split second please be simple functional, you are a post officer and nobody is interest in being in your place, the politicians is the most ugly work on planet earth , only idiot with deep inferioritycomplex, are interest on being politicians Idiot is a Greek word that means one who live in is private dream not connected with the reality or with the universe , that what politicians are idiot , we need tremendous compassion for this ignorant idiot , sorry but to all politicians u have to drop stop to dig graveyard by dealing with weapons, sale buying weapons you have dig a big grave yard for the all world is not allowed is a crime against humanity it seems that you are deaf blind , and you don't know how to read , no sale or buying of death device meant to kill destroy other human being extension of your consciousness is a crime , worst then selling drug much worst , drug nobody is going to force u to buy is up to you , but weapons in a split second you can kill hundreds of people in a split second , is the trade of death you are trading with death , for example Donald Trump as sale 500 billions of dollars of weapons to the Saudi, that are perpetuating a genocide in Yemen , they make air strike in regular base they kill woman children civilians , the number is already to 20000 deaths and there is a spreading of cholera were children are die woman are die every day thank you to Donald Trump the responsible of this genocide, done with weapons made in USA the cash in of 500 billion of Donald trump and the American cost to the world 100000 deaths of civilians children woman, and everybody who is on the path of the air strike , thank you America, and thank you Donald trump , This fashion of the air strike as to be stop immediately dear friend a man who get in a war air craft and throw bomb to the civilians is a monster psychopathic, a beast and fucking crazy , the fashion of air strike as be stop ban abolish by the UN right now , they have being decade were the air strike were not done now it seems that is a fashion enough fucking crazy to throw bomb on innocent civilians population is a crime against humanity, open your eyes to driver of plane say no we are not going , to the captain , refuse the mission , only a crazy neurotic insane man can do that the air strike say no refuse .....Instead in this crazy humanity the one who throw the bomb in air strike is reward with medal is a crazy neurotic insane world or not.....this tradition of wars heritage of the past neurotic humanity as to be abolish ban stop right now, they were unconscious primitive ignorant insane in the past so they make wars , but the world today as evolve in consciousness awareness, civilization intelligence, that the tradition of wars as no place anymore in a new humanity in a new world constitution finish , abolish ban stop not allowed wars bloodshed massacred destruction find something else to enjoy delight, and live the life in a sacred holy way, in peace love compassion, meditation in consciousness awareness, basically in love , make a turn 180 degree from hate unconscious darkness violence, to love light consciousness peace compassion , Stop using different crises around the world as website to sale more weapons , is an old game first you throw coal tar to the windows and in the morning sure is the one that clean the window Donald trump to sale the weapons …… Angelo Aulisa

Published on January 19, 2018 03:25 • 18 views

The new book Mysticism & Physics Chapter 39 Angelo Aulisa

The generation gap part 6

Hi friends , the generation gape , I know it is gap the correct spelling but ape means love in Greek language, so we mix love with the gap because that is what is missing into the world , this post is not about the generation gap , but about actuality , so first NAVANLY president of Russia now , the dictator tyranny of Putin has made countless crime against humanity in the last 4 , 5 years , in Ukraine a massacre invasion and bloodshed , in Syria as gone to help a Criminal of war as Assad that has made uncountable crime against humanity in the last 5 years as kill butchering one and half millions civilians woman children and commune people is own people , with air strike chemical weapons, and conventional war general a real criminal of war he has made countless crime against humanity, Assad as to be brought in international court and charge for his uncountable crime against humanity right now , same Putin he has made air strike campaign in Syria for two and half years bombing civilians killing at least half millions civilians, children woman and population a real murdered and he has systematically siege every city of Syria no water no food no medicine and bombing mercy less , until they surrender a real criminal of war, both this dictator tyranny dictator murdered blood tasting, have to be brought in front of criminal court and charge , now Putin as election he cannot bombing Syria anymore because of the election, but he as teach the job well to is partner dictator Assad , basic is war Stalin style, Stalin he siege the city like San Petersburg because the population of San Peter did not want give up private property and Stalin he siege San peter, and he make starving the local population no water no medicine no food and bombing until the population of San Peter give up private property ,, same those Putin in Syria a real criminal murder Putin is burn internationally and is anti-constitutional to stay in power more than 5 years, he is there since 20 years a real dictator murder criminal of war and countless crime against humanity, we the world call for NAVANLY president right now a young courage man innocent and clean , second in Italy are gone be election soon we the world call for the ban abolish of mafia leader Masson leader Berlusconi that he as ruin Italy totally in his twenty years of ministry he as deliberately ruin Italy ditch Italy into an eternal recession, accumulate a public debt of 2500 billion euro, with an interest of 170 billion per years he has think only to his own interest ,he don't love Italy and Italian people , his govern is 3 minister Masson 3 minister mafia people and two of his own private factory , and is way is corruption and tangent , please an appeal to Italian, band abolish this criminal of Berlusconi now , when he leave is govern Italy Monti take over a disaster of govern, Italy was bankrupt Monti did a miracle to restore Italy account because Monti is a genius , and Renzi as keep the work of rebalance on , Renzi is intelligent innocent young, trust able he love Italy and Italian people and he is not doing is Interest at all Renzi is the one that will lead Italy into the future, into an economic miracle is the only one , those of 5 stelle are crazy totally they know how to say only no , no to life no to celebration no to have good dress no to eat good food no to evolution no to good salary and the salary as to be devolve to the govern totally nut basic, is propaganda is like throwing a spun of sugar into the ocean thinking that the ocean is going to become honey it will remain salty , no to good living life of quality , and is leader Beppe Gillo is a clown, he as an island in south America with big resort and refuge atomic and 4 account off shore and the Italian for him should live starving a crazy man, that every morning he weak up and vomit his unconscious insane full of wounds repression conditioning on up to the Italian population, please to the Italian an appeal Grillo is crazy insane unconscious totally neurotic mad a dangerous for Italian population , do not vote the misery suffering party of 5 stelle otherwise will be the ruin of Italy he will bring Italy to live with the light of candle at the spinning will, avoid to give vote to 5 stelle are the disease of Italy the misery of Italy they know how to say no they are sadomasochist that they like torture people , but they have no program for future they don't propose anything new no future with this miserable party a disgrace of Italy , the new era first of all he copy the terms word from me I am the new era , parrot imbecilely , Renzi is the new era of Italy vote Renzi in Mass please I guaranty for Renzi he is innocent brave intelligent a genius he is the only one that talk correct Italian and he is the one who love Italy and the Italian people the only one that as a passion of love for Italy, the other Grillo hate Italy and Italian people that is why he want starve the Italian people , Berlusconi he is a bandit criminal old style corruption tangent mafia and masonries , and he totally hate Italian people don't get cheated please , this out of compassion for Italian people open your eyes is now or never Vote Renzi in Mass he will lead Italy into the world as leader of Europe and the world generally with great dignity .... Angelo Aulisa

The new book Mysticism & Physics Chapter 40 Angelo Aulisa

The generation gap part 7

Hi friends, the generation gap is suspended I will not write it , if you are interest on it in my 14 books they are few chapter on it very accurate , is useless the world is going the other way around into chaos terrible massacre , war deaths destruction bloodshed all around the globe , all nations are rushing into building produce death device weapons, of all kind nuclear weapons chemical weapons conventional weapons , and war are spreading like wild fire everywhere , life is value is zero zero at the stock market exchange , civilization just a dream an utopia , civilization as never being as never happen in this world , in our ten thousand years of existing civilization as never happen, who knows why people like to talk about civilization I ask myself , ten thousand years ten thousands wars fight , billions of deaths and destruction massacre and bloodshed catastrophe , it seems that human being as being throw into this planet to suffer to be miserable, to kill to destroy to make wars to be in anguish a long story of wars terrible massacre , were there is no war active criminality violence and killing is the Norma , this humanity is a total failure everywhere , I blame the responsibility goes to all religions out of date expire, they are all of them a night mare a bunch of neurosis for humanity all of them are insane and telling lies from the begin to the end of their own culture false pseudo superstition , a total failure of civilization , were anthropomorphism is the disease of all system and nations of the world where they give image character behavior to a fiction call God , they give a face to a faceless reality as consciousness is , and the anthropomorphic man is a real beast, unconscious neurotic crazy, anthropomorphism is a disease worst the cancer believe or not that kill everyday hundreds of people , the neurons of the anthropomorphic man have being totally perverted feed with false notion false information repression conditioning, feed with garbage , and the neurons function into the brain in a chain reaction they are vector of believe thought emotion , and when they have being feed with lies false information and notion, lies and lies , they constitute your own foundation your unconscious , they are like the foundation of your house if they are bogus as they are you will live a false life and soon or later the house will collapse, with bogus foundation they will fall apart , and that what altzaimer is that what a saturated unconscious is , the collapsed of your life it come about around 40 or 42 , when you have spent a bit of life living on lies on false tradition false religions false system false family, is a natural outcome of nature , I am for a total abolish of all false out of date expire religions all around the world, abolish of producing weapons and dealing with weapons all around the world, to produce and dealing with death device call weapons meant to kill an extension of your own consciousness is a crime against humanity , civilization means a world without weapons at first , you cannot bring civilization peace with weapons and wars forget about are you crazy , all the crazy leader dictators of the world they go on talking of peace, producing weapons and wars, is a contradiction in terms you want bring peace with weapons and wars and violence only an idiot can do that, idiot means one who live in his private dream is a Greek word it come from idiom, one who live in his private dream not connected with reality with is inner being and consciousness , that what is an idiot ,, peace cannot be bring about with weapons violence and wars , the first step towards civilization is a world without weapons wars military violence , but this humanity look like that is dump deaf blind , when it comes to actualize the dream of civilization of abolish the weapons band the weapons military and wars , they totally ignore the urgent appeal that the intelligent awakened one of the world is doing in …regular base , is urgent absolutely urgent that the world get this first step toward civilization , look I am not a neurotic in repeat all the time this same music , no first step no civilization that is it , in the past ancient humanity they were use as tradition to fight wars, they were primitive unconscious barbaric , ignorant they live in a dementia senile due the lack of evolution the only think they know was violence fight wars they were totally unconscious neurotic, and that was the tradition of the world , but since then humanity as had evolution convergence of evolution of our DNA quantum leap into the inner being and consciousness and non being and awareness, into the law of eternity core and source of the mystery of the universe and of life and death and of all duality , eternity begin less endless never born or die is the ultimate canvas of the universe were the universe itself is display , that today this humanity as come of age mature, to produce weapons wars military bloodshed massacre is not anymore allowed possible are a crime against humanity , this tradition of wars military producing weapons as to be abolish band completely right now as a crime against humanity, and the factory of weapons that do not comply with the new guide lines of the UN have to be brought in front of criminal court and charge for crime against humanity , or you change item of production beneficial item for humanity, like equipment for reduce pollution and restore the Eco system, then is ok no condemnation or the factory producer of deaths will have to face criminal court . Humanity as to get disconnected from the barbaric past ancient neurotic past, we are not anymore that primitive man , abolish that terrible tradition right now , get update into your consciousness awareness to contemporary that we live 2018 urgent right now , no religions out of date superstition expire no tradition out of date expire , we live here now flowing in consciousness awareness one consciousness for the all universe , consciousness is a fundamental law of the universe intrinsic to the universe just a pulsation of love intelligence a quality a creativity intrinsic to the universe, the very fabric of life , neutral to Gender color race age , no labels no adjectives no contents , the new physic call consciousness Boson x a quality that keep hold the all universe together and confer Mass to particles atoms matter , without the Boson x the all universe and forms of it disintegrate instant , mysticism call the Boson consciousness since thousands of years , is what the ancient old men use to call God , get update , now or never , because the weapons in use today can vaporize our planet in hours , this mad humanity as weapons to destroy our green planet 20000 time, is like you can get kill 20000 tome only of nuclear weapons for not talk of neutron bomb atomic bomb chemical weapons and conventional weapons then is 50000 time, this world can be turn into ash in hours , so the urgency of get an update consciousness awareness , and move on into a new era ha ha, the politicians of the world as idiot parrot, insane, that since I use the term new era they have copy take, they like and they copy me , but my new era is a totally different vision , is a shift from the horizontal line of mind ego unconscious that goes on into an horizontal line from abc to xvz , at an infinite regression , starring at the horizon that tomorrow is gone be the happiness the love but tomorrow never come, what come is always today , now , tomorrow the happiness tomorrow the love the loving approach , tomorrow what come is death that what come starring at the horizon, death is the only unchanging reality into the objectivity all the rest goes on changing mutate , and death is the real communism were we are all the same ,otherwise uniqueness is the criterion everyone is unique into himself , the lotus is a lotus and do not aspire to become a rose , and the rose is a rose and do not aspire to become a lotus ,the rose is beautiful into his own glory early morning with dew drop on the petals and majestic shenth so is the lotus opening is petals to the early morning sun , My new era is a revolution mystical revolution, into the vertical dimension , you close your eyes bring the attention within , the present is the gate less gate to your inner being, the so call dark matter or gravitational energy field will bent the light consciousness deep into your inner being, were you will realize that you are not your thought not your emotion not your body not your sense not your unconscious staff, but you are just a crystal clear witness consciousness, a crystal clear empty mirror that reflect all that surround you and next moment is empty again, those are the chief quality of your inner being witness consciousness and empty mirror no charging no comment , at the deepest point of your inner being pin drop silence is the gate less gate to your inner consciousness, were your inner being annihilate into consciousness, anatta , your inner being melt annihilate into the universal consciousness, this process is trigger by the so call dark energy , or expanding energy field that will pull upward your light consciousness up to the event of the universe up to the horizon of the universe on and on with the same expansion of the universe, you in essence will expand , intrinsic to the universal body , actually you in essence are the universal body itself, like I was waiting for the rain this morning for weak up and in time the rain has come and I weak up , a joke to say that at this stage you are the universe itself the same size and big the same , a breakthrough sacred divine immense psychedelic ecstatic a bliss , and on and on the dark flux or flowing energy field will move circulate your light consciousness in an overlapping with non being body in a cosmic play irrational , you in essence breath, the universal body breath into non being body and non being body breath into the universal body that the play of the flowing energy field, then nothingness emptiness are the gate less gate to non being body, were forms time space duality, totally annihilate and consciousness also annihilate because consciousness exist always in relation to a subject or object, when the object and subject annihilate also consciousness annihilate or better it twisted into unfocused awareness , ultimate essence of non being body, awareness is just an I am ness relation less unfocused because into non being no forms no time no space no duality, all is transcended annihilated , you know you can call meditation annihilation process , here you in essence are at home , non being body is infinite, the size is infinity the universal body annihilate vanish into infinity, is just an open relativity infinite, an infinite freedom an infinite peace an infinite silence rich of subtle ecstasy an infinite bliss sacred divine, here immortality is rediscovered here the resurrection happen, even if you still hold a body the resurrection happen into the body, mean while you are in the body, actually is call enlightenment, your unconscious sleep gone forever you in essence are awake enlightened forever and ever , immortal resurrected if in this alchemy you don't hold a body you in essence are resurrected body less, for eternity to come body less in essence fragrance data of intelligence that you have refine in your life, you will flow into eternity into awareness, and into the universal body into consciousness, extend into our objectivity everywhere and nowhere in particular expand into the universal body, into the living being into the air, into the wind, into the nature into the animals layer, into the universal body generally intrinsic to it as an essence fragrance data of intelligence of it you will be roses you like it ha ha, you will the oceans trigger ing new , waves ha ha new living being each living being is like a wave of the ocean that rise and disappear at its own turn into the oceans, what remain is the ocean the oceanic consciousness , that is the new era vertical , a full circle like a zero that what your consciousness and DNA aspire for a complete realization , actually is an eternal journey that end nowhere life is eternal , endless begin less as eternity is and then you have accomplish your journey the journey of your consciousness totally refine into light empty consciousness, no more unconscious asleep just a pillar of light empty consciousness , this is the new era the new dawn of consciousness awareness, this world the lotus paradise this body the very awareness consciousness . The situation of the world is very chaotic neurotic a chaos, but out of chaos stars are born out of dust gas debris stars are born, so don't be scare is good but just take the right turn within your inner being vertical, and is done ,each one individually as to be decisive determinate to apply this process of awakening enlightenment is a revolution that you are decisive about , if you let go into evolution it will not happen a natural evolution do not bring about the parcel of enlightenment ,your effort dedication is require totally at the begin then it come effortless , all the search all the method of search are to tire your mind ego unconscious to exhaust your unconscious mind , then you drop all the method and you let go into enlightenment , effortless , then bliss ecstasy silence, peace then eternal life, death is the greatest fiction of humanity it do not exist, then you are in for the greatest surprise of your life, you in essence are eternity itself the journey is from eternity to eternity from here to here endless eternal , I could stay here talking forever to you all but I guess is enough for today ...Angelo Aulisa

The new book Mysticism & Physics Chapter 41 Angelo Aulisa

Toward civilization

Hi friends , we have to remark the score of two big point of China , so China is talking a different language then the insane unconscious neurotic Americans and their president , unconscious Donald Trump , first of all when they meet Trump and the president of China few months ago , Trump he support an old ancient mentality of patriotism and America first like an old record broken and stark there , Trump is the divided he split the world in thousands of fragment of patriotic nations , old fashion fascist . Trump he split the world in dealing the trade market he as a mentality to come back at the stone age wen was no computer no web no communication simultaneous no telephone no television no radio no flights a barbaric primitive man fascist , instead the president of China say the world is a global village and he support a global market a contemporary man of our age , 2018 , a wise man , then about nuclear weapons and chemical weapons and atomic weapons neutron weapons that are dangerous like hell a crime against humanity that can wash away the world in hours , Trump as the same old mentality of 70 years ago when it was cold war Trump he talk about produce new nuclear missile small nuclear missile to confront the Russia is a crime against humanity as is not allowed ban abolish by the new guide lines of the UN, to produce more nuclear weapons as the world as show with the Nobel prize for peace given to I CAN that is an association for band abolish deactivate all nuclear weapons and deadly weapons of mass destruction all over the world , but Trump blind deaf ignorant primitive unconscious, he ignore the urgency of the world to see a world without this terrible menace of nuclear weapons , and weapons of mass destruction , he play Trump the same ugly game of cold war an old mentality to throw coal tar to the window of the media, in order to sale weapons all over the world he divide nation putting one against the other, so he can go later on to sale weapons a dangerous for all humanity that what is this Trump , instead the China talk about old mentality of cold war and we should instead talk of disarmament deactivation of all nuclear and deadly weapons great thank you , the president of China is showing much more wisdom intelligence he is a wise man a ray of hope for the world , dear Trump the equilibrium the harmony the balance between nations as to be keep by intelligence consciousness awareness is finish the age of old mentality of cold war that the one nation who as the most deadly nuclear missile his the most strong nation of the world , the law of the jungle and of the insane neurotic old little men is over , the world as evolve as had convergence of evolution of DNA and quantum leap on consciousness awareness meditation , that today days the confrontation between nations is done through intelligence consciousness awareness ,conversation discussion , dear Trump if you go to Putin and tell him let talk of disarmament deactivation of nuclear weapons , and tell him we need to disarm deactivate a few strategic position of nuclear weapons simultaneous into the world , we the American we will deactivate disarm this nuclear position into the world simultaneous with you Russian 7 , 8 position base of nuclear weapons , the answer from Putin will be YES ok let talk about which position strategic base disarm deactivate , instead of doing as you are doing of building new nuclear small missile , that old fashion as never function work , please Trump even if you are blind deaf ignorant , try to be a little smart as you say , you don't need to build new missile but talk with Putin of disarm deactivate a few position of nuclear missile , Putin will agree Trump please stop your policy of making money through selling weapons all over the world through using crisis as website to sale your fucking weapons that kill millions of people around the world, that dark age mentality is over come to conversation peace table with the other nations the answer will be yes , even north Korea are very innocent people you record nice there status of nuclear nation and call them for peace deal and talk and disarmament the answer will be yes , dear Trump the world is a small global village that has come together in my understand you are a dangerous for the all world the nuclear weapons in your hands are not safe because you are old serious fashion ,you never evolve and you think only about money having more money dealing and sale weapons all around the world is NOT ALLOWED ANYMORE IS A CRIME AGAINST HUMANITY to kill an extension of your own consciousness each living being is sacred holy , we live in an organism call universe and we are one through consciousness, weak up man out compassion for you and all the world weak up , go for peace deal and conversation of disarmament and deactivation of the worst menace that the world as ever had the nuclear weapons and chemical weapons neutron atomic bomb etcc... right bow follow the policy of the China that they call for conversation discussion of disarmament , for me you are a dangerous for all humanity, Trump nobody knows how you come elected as president maybe the American were drunk that day of election , but there you are now you deal on peace with peaceful meaning ok , America is not a civilize country you need to wear a weapons to go around your society as if everybody is your enemy is not so civilization means common sense that everybody is a friend a beloved friend that is civilization and you don't have health care any disease in America and they let you die in the street , America have no consciousness even on Eco system they don't care that the world is on the verge of self-destruction through calamity , so I put America as the most not civilize country of the world , were thousands of ton of cocaine are consume everyday were everybody almost is a drunkard , were they consume chemical medicine of opium that is almost like heroin that give addiction and in days, and then you need to consume in regular base otherwise you feel bad , and die maybe , you American have the biggest population of obsess ugly like hell you don't know how to eat , and you glue to television and computer 12 hours a day is dangerous for psychological health and cancer the outcome and funny you American want teach to the world democracy, you don't have in your own country how you can teach , please come to your sense to the American ask the resignation of Trump, and begin fresh a new begin with a sane president this choice of yours of Trump is a dangerous for all the planet, he cannot hold in his hand the button of nuclear weapons is not psychological feet , , and please follow the direction orientation of the world of disarmament deactivation of nuclear weapons and weapons of mass destruction follow the direction of the China that they talk of this new dawn of understand , instead of building new nuclear missile talk with Putin he will answer Yes to disarmament Russian people are intelligent people ....Angelo Aulisa

Published on February 05, 2018 01:20 • 22 views

The new book Mysticism & Physics Chapter 42 Angelo Aulisa

Toward civilization part two

Hi friends, a tale a story, they was once and onetime a great Sufi master that was very well known notorious, call Kidr.. And as all Sufi master is ruff was the sky, and was going around from village to village , and every village was waiting for Kidr the famous Sufi master , he arrive in a village call Goutha , and all the villagers were very happy , they arrange for him a beautiful room to stay and all the village was waiting for his first sermon discourse , In the morning after early all the villagers they arrange make a beautiful chair under a big tree for Kidr to speak , and all the village was getting there to hear the great discourse , The beloved Kidr come seat on the chair all the villagers thousands of people were there , and pin drop silence , Kidr was silent too for few minute then he say how many of you knows what I am going to say rise the hands , and all the villagers rise the hands , then Kidr weak up and say then there is no need to speak if all know what I am going to say there is no need to say anything and went away , now all the villagers were upset disappoint , so the head of the village went to Kidr and say sorry please come to deliver your sermon we apolligiaze , Kidr say ok tomorrow morning I will deliver the discourse , the morning after come , and Kidr went under the tree and seat on the chair silent , all the village was there anxious to hear and same Kidr say how many of you knows what I am going to say rise the hand half of the congregation rise the hand and half remain stand still , then Kidr say the half of you who knows what I am going to say tell to other half and weak up from the chair and went away , all the villagers were frustrate disappoint , and same the head of the villager went to Kidr and say please sorry come to deliver the sermon , Kidr out of compassion say ok tomorrow morning I will come again , the morning after come and same Kidr Arrive and silent he seat on the chair arrange for him , it was pin drop silent for few minute and then Kidr say how many of you knows what I am going to say rise the hands , this time nobody rise the hands and it was pin drop silent , then Kidr say if nobody knows what I am going to say there is no need to speak you will never understand , and weak up and went away , all the villagers were pist off upset frustrate , same the head of the village went to Kidr and this time pray Kidr in tears to speak to deliver the sermon , the morning after of course Kidr was there on the place arrange for him silent and all the villagers and the near villagers were there absolutely anxious to hear the discourse of Kidr this time he deliver the greatest mystic discourse ever hear in the universe , and all the villagers were bliss in ecstasy happy , then the morning after Kidr went again and deliver the same identical discourse of the day before , the villager were surprise but they do not say anything the master was so great out of respect they don't say anything , the day after again Kidr deliver the same identical discourse at which at the end the head of the village weak up and say in front of all the villagers sorry master but this discourse is already the third day that you repeat the same identical discourse at which Kidr start to laugh loud for minute until tears fall from his eyes , and say I know but I have being watch you the last 3 days and I have seen nobody of you actualize what I say , I have tell you to meditate for real and nobody meditate so until you actualize the first step of my discourse that is meditation to meditate for real I will go on repeat the same discourse there is no need to say further , he was a majestic Sufi master the greatest ever walk on this earth , and same I say to you all friends you have not even actualize the first step of the discourse that I say which is the ban abolish of all the weapons all over the world nuclear atomic neutron chemical weapons , the stop ban abolish of production of all weapons all over the world the factory industry that produce such a device of death have to be brought in front of criminal court and charge for crime against humanity is a crime to build device of death call weapons that kill an extension of your consciousness a living being sacred holy divine , one consciousness for all organism of the universe , until this first step toward civilization is actualize is useless and a waste of energy to go on talking to you , the world is going the other way around off civilization into wars tragedy bloodshed destruction deaths , they like this they are megalomaniac dictators like Assad , Putin , , Erdogan that they have the deepest inferiority complex a dangerous disease and dozens of others dictators that are absolutely blood taste murder criminal of war, they have commit countless crime against humanity and this humanity is doing nothing to remove this monster megalomaniac murders , so for what to go,, on speak to you , I am not a politician ,or I am not interest in politics , and I am harmless nonviolent man , and powerless, in this world civilization will remain a beautiful dream freedom will remain a beautiful dream peace will remain a beautiful dream love will remain a beautiful dream enlightenment an utopia , for you all , is useless and wastage of time energy , this tyranny dictators prove to be the greatest megalomaniac neurotic insane men on earth with great inferiority complex a serious disease fascist patriotic idiot , we will have to wait years that they will be throw into the garbage , when democracy meritocracy will be restore again , we can speak again ok ,for example the military parade is a prove that Trump is a megalomaniac traditional old fashion little serious man insane neurotic , friends this Trump as serious psychological problem into is unconscious is totally unconscious identify with out of date expire tradition death tradition he is the biggest hindrance for the evolution of humanity together with the Muslim religion they block hinder the evolution of the world at large they bring a dark age on like it was hundreds of years ago no joking , by the way the air strike have to be band abolish stop into this world right now , they have being decade that they were not done band , now they are in fashion , each nation who use air strike as to be brought in criminal court for crime against humanity right now , are the vicious criminal action of this megalomaniac dictators tyranny imbecilely , insane neurotic monsters no other definition , no wars no military no weapons band stop abolish forever from the new UN constitution , enough , of assassin murdered tyrant you have to pay to humanity your countless crime against humanity one day you will pay ......Angelo Aulisa

The new book Mysticism & Physics Chapter 43 Angelo Aulisa

Toward civilization part 3

Hi friends , believe is for the ignorant , one believe when he do not know to hide is ignorance, because nobody like to feel that is ignorant he put on a mask of believe , mask is a Greek word that means persona in fact believe generate a persona an ignorant that hide behind the mask , and believe generate the disease of anthropomorphism when you give an image character behavior to God you personalize God, you think that God is a person , you give a face to a faceless reality as consciousness his , anthropomorphism is a disease worst then cancer that kill everyday hundreds of people , and make man crippled spiritually retarded, in fact this humanity is spiritually conscious retarded and make men superstition in fact this humanity live in the worst superstition prejudice out of date expire , and believe in religions out of date expire generate conflict antagonism wars bloodshed massacre destruction deaths, between the different out of date expire religions, that preach only lies and lies insane neurosis from the begin to the end of their philosophy, and this is the biggest cause of killing deaths destruction massacre and bloodshed into the world today , believe generate short sight you see up to your nose , believe generate perverted neurons your neurons when you believe are feed with lies and lies false notion false information conditioning repression wounds , and your neurons function into your brain as chain reaction of believe emotion thoughts , and when have being perverted by lies wrong information notion conditioning of false religions that preach fairy tale for retarded children, lies and lies they constitute the foundation of your unconscious life, of your house and with bogus foundation your house your life sooner or later around 40 or 42 , will collapse tell apart and the consequence are altzaimer saturated unconscious that generated dozens of psychological disease and stop of learning the cell of learning of your DNA die , and consequence also cancer because the majority of cancer happen when you go against nature freedom spontaneity naturalness , . Instead for me trust and experience and experimentation is the way the path , when you experience your inner reality your inner being when you realize your inner being through experience of meditation love silence , believe stop drop dead because now you know trust , believe is transform into trust through the experience and you trust because you know your inner being , do you believe in the sun no you trust the sun because you see the sun right in front of you , same is your inner being soul , you know it through the experience of meditation you have known it melt annihilated into it and you are it your inner being, which is rediscover in the vertical dimension of deeper and higher of the present the moment here now flowing into the present moment , and non in the horizontal dimension of ego mind unconscious , believe is for the ignorant bigot that live in the horizontal dimension of ego mind unconscious , trust is for the intelligent for the new men that live in the vertical dimension of deeper and higher of inner being of consciousness of non being of awareness of the law of eternity core and source of the mystery of the universe and of life and death and of all duality , were time space forms duality totally annihilate, and what is left is awareness unfocused awareness that is the ultimate essence of non being awareness is relation less just an I am ness everywhere and nowhere in particular expand into non being incorporeal body, into eternity begin less endless never born or die , eternity is the ultimate canvas of the universal body , huge bigger vaster then the universe itself actually the universal body is display into the ultimate canvas of eternity , and into non being immortality resurrection are rediscover happen , that is the aspiration of your DNA and of your consciousness that are intrinsic link interconnected consciousness and DNA . And is the triumph of human being on earth of civilization the higher apex of civilization is to realize that you in essence are eternity itself intrinsic to it one with it in short you are it eternity itself infinite no size unbounded , infinite freedom infinite silence rich with subtle ecstasy , infinite bliss sacred divine , infinite peace infinite intelligence because eternity is the ultimate source of intelligence , infinite love because eternity is love pure love unconditional beyond duality , is just an open relativity endless not absolute at all because it end or begin mo were , is an eternal journey infinite like a circle like zero , from eternity to eternity from here to here it start from eternity unconscious and it end into eternity totally conscious your consciousness as refine into light consciousness totally refine into light consciousness and it return to the source into a circle of zero , .resurrected enlightened totally awakened then the journey end into eternity, annihilate into eternity, and is a new begin body less formless in essence fragrance data intelligence, intrinsic to eternity for eternity to come you will flow intrinsic to eternity, and to awareness, and to the universal body, you in essence will have a universal body, formless and intrinsic to universal consciousness extend to the objectivity, into intrinsic to all forms of the universe living being animals nature wind oceans stars planets galaxies, you in essence are everywhere and nowhere in particular expand into the universal body and eternity itself, like l Lucy but without any drug ha ha ha . The American are a population of believers but in weapons in wars in bloodshed in violence , in killing the other because basic for them the other is the enemy , not the fiend not the beloved friend the American are not a civilize country I repeat until they will believe in weapons in wars in military in killing and violence , look I have nothing against the military are sacred holy living being as any other forms of life , but only are an heritage of the past ancient humanity wen men was living primitive barbaric neurotic insane due the lack of evolution in consciousness awareness and DNA , humanity as use as tradition to fight wars, they have nothing better to do , they were ignorant to interact through wars killing destruction conquering this and that , but today human being have come of age mature thousands of convergence of evolution into our DNA and thousands of quantum leap in meditation consciousness awareness that humanity as grow evolve immense in intelligence consciousness awareness, that we are one into consciousness an oceanic universal consciousness for all universe just a fundamental law of the universe a pulsation of love intelligence of creativity, just a quality diffuse intrinsic to the universe everywhere and nowhere in particular expand into the universe one for all universe , that fight wars killing bloodshed are not anymore possible not allowed to kill an extension of your consciousness a living being sacred holy divine , is a crime against humanity , I have nothing against the military no condemnation are nice fellow only the military are a regrettable heritage of the past ancient humanity that we have to get reed off , the military can be transformed in thousands of other job work , no wars no weapons no military abolish band forever that is the first step of civilization no first step no civilization , a world without weapons is a priority of civilization , this planet the lotus paradise this body the very consciousness awareness , that is it for civilization to flower , Men can face alone the sunrise of consciousness awareness we don't need mediators of fake religions fake priest that they have preach lies and lies all the way along , man alone in silence can face the new sunrise of consciousness awareness is enough into himself actually is a light into himself ok.......Angelo Aulisa

The new book Mysticism & physics Chapter 44 Angelo Aulisa

Toward civilization part 4

Hi friends , it may be useless a wastage of time , but maybe it may help to understand the situation right now , is always the same when you live in an unconscious system society , were people they have not have any alternative to become conscious aware through academy mystery school, were they learn meditation and consequence there inner reality , of rediscover there inner being realize there inner being which in essence is love, compassion peace bliss silence , and the link bridge to their universal consciousness , non being incorporeal body, awareness, into the law of eternity core and source of the mystery of the universe and of life and death and of all duality , that is what make an individual conscious aware , the experience of meditation done for real , and as far as I know they exist not such academy or mystery school very few here and there in the world very few , this alchemy process of meditation is a shift a quantum leap from the horizontal dimension of ego mind unconscious, that goes on horizontal linear from abc, xvz, at infinite regression , tomorrow is the happiness tomorrow is the love happening but it remain always at the horizon and the horizon goes on receding back, it remain always an illusion it remain always at the horizon the happiness the love, tomorrow never come what come is today what come is always now the moment by moment the present , and into the present gate less gate to your inner reality inner being, happen the shift quantum leap of the vertical dimension of deeper and higher , you close your eyes bring the attention within, the so called dark matter or gravitational energy field will bent the light consciousness deep within your inner being, were you will realize that you are not your body not your mind not your thought not your emotion not your unconscious staff not your senses, but you are just a crystal clear witness consciousness, mirror like an empty mirror that reflect all and everything that surround you in the moment and next moment the mirror is empty again , no comment no charging , this are the chief quality of your inner being witness consciousness and being an empty mirror Belgium mirror the best quality of mirror , and the essence of your inner being are love you into your inner being are love pure love, compassion bliss sacred divine , intelligence actually the source of intelligence is your inner being and not locate into your brain into your brain is locate the intellect , but into your inner being is the source of intelligence love bliss compassion , peace silence , and deeper into your inner being, pin drop silence gate less gate to your inner universal consciousness were the inner being annihilate into consciousness ,and you in essence are just universal consciousness expanding at infinite into the universe the so call dark energy or expansive energy field will pull upward your light consciousness is what trigger your expansion upward into the universe at the horizon of the universe at the event of the universe , your size become the size of the universe expanding on and on and you in essence are intrinsic to the universe and big as the universe itself, is a break through sacred holy divine ecstatic psychedelic majestic , the universal consciousness is just a fundamental law intrinsic to the universe the very fabric of life, is just a pulsation of love intelligence just a quality a creativity expand everywhere and nowhere in particular into the universal body, and you in essence are just intrinsic to the universal consciousness in short you are actually it universal consciousness expanding with the universe on and on , and then the so call dark flux or flowing energy field will move flow circulate your light consciousness in and out into non being body incorporeal body, into an universal breath , the universal body that you are in essence will breath into non being body and non being body will breath into the universal body overlapping each other, at this stage nothingness emptiness are the gate less gate into non being body, were time space forms duality annihilate completely, and consciousness too annihilate into non being body because consciousness is always in relation to a subject or object wen no more object or subject are there consciousness cease to be annihilate into non being body incorporeal, or better it twisted into unfocused awareness relation less awareness that is the ultimate essence of non being body, and just an I am ness a great relaxation , even in ordinary life you become conscious always about an object or subject when the subject or object is dispose you twisted into unfocused awareness relation less awareness ultimate essence of non being body incorporeal , is a continuous of being conscious and twisted into awareness were a great relaxation happen no more relation no more focused of object or subject and you relax back into the core and source of non being body were no time no forms no space no duality exist and awareness is just an I am ness expand everywhere and nowhere in particular into the law of eternity, which as no begin or end never born or die eternity is the ultimate canvas of the universal body, were actually the universe is display and eternity is huger bigger vaster then the universe itself, eternity is the core and source of the mystery of the universe and of life and death and of all duality , here you in essence are at home the size of eternity vanish annihilate into infinity your size now is infinity infinity the ultimate equation of quantum mathematics , the core and source of eternity is infinite freedom , infinite bliss , infinite silence rich with subtle ecstasy , infinite intelligence actually the real source of intelligence , an infinite open relativity and not absolute at all because it end nowhere or begin nowhere , you at home into the core and source of the mystery of the universe , and here the realization of immortality the realization of resurrection happen, and the unconscious sleep gone forever you are awake enlightened forever and ever , the resurrection is a conscious alchemy from unconscious to consciousness awareness into the law of eternity not physical not gross not material but a conscious alchemy , This is where the process of meditation done for real can lead an individual , but the basic point to focused is that the unconscious sleep is gone forever and ever , that the path to consciousness awareness , and if every individual as the alternative to experience this path I guaranty that in essence a human being is pure love unconditional beyond duality , is pure intelligence pure compassion pure bliss beyond duality , unconditional beyond duality , and is unconscious is transformed into a pillar of light consciousness . Now the point is that all the system of the world society of the world are absolutely unconscious and live into the mind ego unconscious , and the unconscious wen is saturated to much it stem up like pot boiling , and wen is too much really saturated it can happen a pent up suddenly one is in the oblio of the unconscious not knowing anymore what is doing literally in an outrage wild the thin conscious totally erase wash away , and then one is dangerous very dangerous , but himself is not at all aware or conscious of what is doing is in a rage wild and dangerous , hence the importance to keep always an unconscious flowing and free of condition repression wounds trauma , and the path is meditation or psychoanalysis , but psychoanalysis is very limited it can make the patient temporary heal for hours after the session the patient as a relieve but then he start accumulated again dust into the unconscious , meditation instead make the patient totally heal because the priority of meditation is to center the patient into is inner being, that is at a certain distance from body mind thought emotion senses unconscious staff ,your inner being encompasses your organic unity at a certain distance , so even if unconscious staff is there or emotion or thought or ego mind you watch at them from a certain distance and the watch the witnessing ,,,,,,,,, process burn the wounds off and the cosmos reabsorb them and you are lighter forever, in the end of this process of meditation after years your unconscious is transformed into a pillar of light consciousness and that what is enlightenment bringing light into your unconscious more light and more light until no dark corner is left until your unconscious is just a pillar of light consciousness, then you are just light consciousness in mystical union harmony balance with the universal body and eternity itself , the process of meditation heal completely an individual and give to the patient individual strong roots into his own inner being into his own inner consciousness into non being incorporeal body into awareness into the law of eternity at home in peace into an infinite relaxation then is done the unconscious sleep totally gone the individual is just love peace compassion silence intelligence beyond duality unconditional immortal resurrected . I am sorry to see the American in such difficulty first step take away the weapons from the hand of the people all the weapons unconditionally, a weapons in the hand of an unconscious man sooner or later he will use it and create a massacre , any discussion controversy and the unconscious can bobble up , is dangerous like hell , the unconscious is nine tenth and the thin conscious is one tenth , when it bobble up it knows no reason it wash away the thin conscious in a split second , so abolish band stop of selling weapons , is a crime against humanity to produce device of death that can kill living being sacred divine and extension of your own consciousness , change production in beneficial item for humanity if is a matter of jobs working and making money, enough is enough they are so many other item equipment to prevent pollution for example but thousands of beneficial item for humanity , American when they go out in the society they have to understand that everybody is a friend a beloved friend and not an enemy that is common sense civilization and instead of wear a weapons they should wear love a big smile friendliness compassion that ,what civilization is all about being lovely compassionate friendly , please abolish band the trade of death device call weapons now at least in your system society , is too dangerous in a society that live totally unconscious high risk every second out of love for all American my grandmother was born in Washington I love American people ......Angelo Aulisa

The new book Mysticism & Physics Chapter 45 Angelo Aulisa

Toward civilization part 5

Hi friends , again it maybe a wastage of time useless but it may help someone , let go deep into the foundation of the unconscious again and again , a human being start to build the individual unconscious already before that he actually born , into the mother womb , the behavior of the mother the attitude of the mother the character of the mother the situation in which the mother live affect the child already into the womb , then the child born and since the begin of is life already he start to get build up the unconscious , the child in the early age of is life since the first months is absolutely helpless , the family goes projecting on him all their culture behavior character , the child absolutely unconscious absorb all and everything totally ,and the most sensitive age is really the early age of is life from zero to seven years the child build up the foundation of is l unconscious and life , is neurons are feed with all the knowledge of the father and mother , that means wrong information wrong notion of life as system society tradition , of culture and out of date expire religions lies and lies neurosis and conditioning repression of all kind wounds and trauma start to build up too , that means that 99 per cent of couple today live a situation of family of domestic violence abuse fight all the time, the child innocent absorb all the vibration of those insane drama etc... so the neurons of the child which are the foundation of the unconscious, get loaded with all this unconscious staff of his family , the neurons once feed whit all this conditioning repression lies false notion false information of family, society and false religions get completely perverted , and the neurons function as vector in a chain reaction into the brain of thought believe emotion ,and once are wrong feed they constitute the foundation of your life and your life will be pseudo false forever and ever , you will live a life of the other who as projected the unconscious on you , the unconscious do not belong to you , but to the society system were you are brought up , and psychologist say that at the age of seven almost all the path of your life is determinate in print , actually psychologist say that men stop to grow to learn at the age of seven wen the foundation of your unconscious are lied down , and sure enough true enough your life will follow the path that your unconscious as being programmed for , without exception , so be careful what you projected into your child up to seven years of age that will determinate the life of your child , after seven up to 14 still is a very sensitive age to build up the unconscious of the child and the child goes on accumulate all kind of information notion into is neurons that are equivalent to your unconscious , the in print goes on and at the age of 14 let say the neurons are totally completely feed the unconscious as being projected into you , and sure enough you will follow that path and you will live the life of others of out sources of society system, false religions unconscious absolutely not aware not conscious , your parents are helpless and not responsible of doing this disaster they too are absolutely unconscious follow the path of their unconscious and of the collective unconscious of the situation where they live , the collective unconscious enclose in it all tradition of false religions false system society all kind of conditioning repression lies wounds trauma of the system society they belong , an Indus will grow as an Indus a Christian as Christian etc.,,, they are dozens of different collective unconscious and are very powerful conditioning on you that rare you can escape almost impossible , and so the child at the age of 14 as already all the blue print in print for all is life , and it goes on accumulate is unconscious every second on and on in all is life that why the unconscious is nine tenth and the thin conscious is only one tenth no comparison the unconscious is the all mountain and wen ever a situation will face you sooner or later the unconscious will assert itself with revenge and violence disaster will erupt up , and so one goes on living with bogus wrong false foundation all is life , when you reach the age of 40 or 42 that you have spent already a beet of your life many inner question will bobble up , say Carl Gustav Jung true at the age of 40 or 42 you start questioning yourself who I am, were I come from, were I am going what about death and life, the mystical inner question that everybody has , they start pinch you from within the inner nature is pushing for answer , and that is the time when dramatic disease break out like altzaimer and dozens of psychological disease , because your foundation of your neurons of your unconscious are bogus false the foundation of your life are false your house your life will collapse tell apart , disintegrate and you find yourself into an abyss of void in ordinary sense and you start to fail into the abyss, frustration anxiety anguish is the outcome suffering misery etc.... I always suggest to live the neurons of the child the unconscious of the child empty, free of conditioning repression wounds trauma , what you know you tell to the child what you don't know be sincere you tell to the child we don't know you will learn by growing you will chose by experience by growing , like this the child will be stimulated to question life to inquire life to explore life and he will come up with is original conclusion synthesis , he will be an agnostic one who question explore life and make is conclusion original through is experience he will get is original wisdom , a joke once a child was ask what is your name the child answer don't the questioner answer are you sure the child yes my mother and father always tell don't do this and that all the time so my name is don't , do not do this please . To this terrible disease of the unconscious because is a disease that determinate the death of the cell of learning into your DNA it will generate the disease of anthropomorphism that you give an image character behavior to God actually all humanity think that God is an old wise man with white long bird seating into a golden throne into the sky with a magic stick , is dramatic but is like this anthropomorphism give a face to a faceless reality as consciousness his , and the unconscious will generate cancer because the majority of cancer is generate by going against nature freedom spontaneity the repression of freedom nature spontaneity the repression of sex, which society religions false, celibate is impose that create the disease call cancer, the cells start to die because they find no out let of expression, and the cancer is generate is the nature of our DNA wen happen a convergence of evolution if you welcome it the evolution take place if you shrink the death of the cells that were ready to evolve. Die and cancer is generate. I am showing to humanity an alternative of approaching life from within first and without objective, to a new dawn of meditation intelligence civilization inner being consciousness awareness, that will change the path of humanity completely totally, and is the only true authentic alternative this inner revolution need academy mystery school millions around the world were people will have the alternative choice to decide their path , original from their own inner being link bridge from the known to the unknown to the unknowable to universal consciousness, a fundamental law intrinsic to the universe, no labels no adjectives no contents no interpretation of the little serious men , consciousness is to infinite to define in labels adjectives contents , and neutral to Gender color race age , is just a pulsation of intelligence love just a quality a creativity intrinsic to the universe, and you can be one whit it through meditation silence love done for real , and on and on into non being body were forms space time duality totally annihilate, and awareness unfocused relation less is the ultimate essence of non being body, awareness is everywhere and nowhere in particular expand into non being body incorporeal is just an I am ness diffuse into the law of eternity, is just a great relaxation into home into eternity itself , were the unconscious sleep is gone forever and immortality and resurrection are rediscover happen , so the only way forward is a world conscious aware the only direction orientation were this planet the lotus paradise and this body the very consciousness awareness for a new era to happen to unfold .....Angelo Aulisa

The new book Mysticism & Physics Chapter 46 Angelo Aulisa

Toward civilization part 6

Hi friends , again it maybe a wastage of time and useless a wastage of energy , but still it may help someone , again let go into the unconscious , I want say a maximum you never being free a single second in your life, and freedom is your inner nature your inner essence hinder the freedom and you have already create the path to build a teak unconscious , for example you have born accidental in your country you never chose your nationality and you have being impose a religion in base of the contest you belong ,you never chose your religion as being impose on you as being an accident , this fact show you that you from the early begin of your life you have never ever being free , no freedom since ever to choose decide your own path , as being a simple accident you could have being born some were else and your life would have being totally different for example , hence my insistent to live the neurons of an individual free of conditioning repression wounds, wrong notion wrong information, empty , so then consciousness is the feed of your neurons and the intrinsic quality of consciousness are intelligence , love , freedom , bliss sacred holy divine , silence rich with subtle ecstasy , compassion , playfulness , celebration , once your neurons are empty consciousness will be the contents of your neurons, and the outcome is a life of truth sacred holy divine a life of love of peace of silence of ecstasy a life of bliss a life of freedom of celebration of playfulness , with deep roots into your inner being at first link bridge to the unknown and the unknowable of universal consciousness, non being incorporeal body, were time space forms duality totally annihilate and awareness unfocused relation less is the ultimate reality, awareness is just an I am ness diffuse everywhere and nowhere in particular into the law of eternity, awareness is just a great relaxation into home into the core and source of the mystery of the universe and of life and death and of all duality , you are relax at home into freedom infinite, bliss sacred divine infinite silence, into infinity the size of eternity vanish into infinity just an open relativity endless begin less infinite freedom , wen consciousness is the content of your neurons your life is transformed into a divine sacred bliss, peace silence, and eternal, immortality and resurrection are intrinsic to your neurons , that give to you absolute relaxation when you know death as a fiction, the body die but your inner consciousness is eternal never born or die is eternal , until you know death as fiction you cannot relax , once you know death as fiction by experience then a total complete relaxation , a new fresh begin only then you will live your life as a sacred holy cosmic play, and enjoy as never ever, relax in peace at home . what to do say Freud a man cannot be lived wild without any education it as to be give an education otherwise man will be a wild beast , but well known that the education come at heavy price the end of freedom, and this create the building of the unconscious the feed of your neurons whit fancy tale neurosis lies of religions of system society ,family , and then we need dozens of psychologist to readjust your neurons unconscious is this not a mad game , my understand of education is to unfold from within outwards and not vice versa as it is in our society today that is unfold from without to within , if this is understood many almost all disease consequence are avoid , education means unfold from within your inner being out help your inner nature to express itself outwards without interference of outside sources, of society religions false and system false and family false neurotic , we just help the individual to unfold out is inner nature quality attitude from his own inner being , this will change the all picture system society family , nothing should be impose on the neurons unconscious of the individual , then we will have a totally different approach to life either from within that from without , almost all of the psychological disease neurosis will disappear . And once the individual is left in freedom whit is neurons empty free he can chose is path in freedom, whatever he feel for he like, he can chose is nationality the universe I guess, he can chose is religion less religion his consciousness I guess, that is what freedom is, for the first time we will have a humanity that live in freedom for real with deep roots into is inner being into is universal consciousness non being awareness into the law of eternity begin less endless never born or die, that the meaning of eternity , the core and source of the mystery of the universe , that will be a new dawn of meditation civilization consciousness awareness into the law of eternity, ultimate canvas of the mystery of the universe were actually the universe id display , freedom is the magic word, to avoid a loaded unconscious, remember it always do not hinder the freedom of anyone never ever is a crime that create unconscious, loaded neurons and disasters , saying by the greatest psychologist like Freud Carl Gustav Jung Adler , and the greatest mystic of the world the greatest mystic master of the world, keep it in consciousness freedom the magic world , no freedom no love, freedom and love are deeply interconnected related always and always synonymous ......Angelo Aulisa

Published on February 16, 2018 05:40 • 24 views

The new book Mysticism & Physics Chapter 47 Angelo Aulisa

Towards civilization part 7

Hi Friends , we continue even if I know well that is wastage of time energy and useless, because I have no power of any kind I am a commune man, nonviolent harmless, a mystique, one for whom meditation his is life path , well one thing very important to know about the world of psychologist and psychoanalysis and consequence ego unconscious mind is this , that wen we are working with the unconscious of a person and we find conditioning wounds repression trauma , and we bring those on the surface of social life, actually wen the patient denied it strong violent or with obsessive insistent is because the wounds is there hundred per cent, otherwise the patient interaction is absolutely calm indifference or laughing about the accusation , accurate it function like this , for example the insistent obsession of Donald Trump to say all the time no collusion like a record broken or no meddling and even say that the police was too busy to look after the collusion so they don't check out other story, is a turning around of people so hypocrisy that is a prove of is guilty in the case , how the police can check out 300 million people , is insisting in saying the contrary obsessive make him guilty, and so vicious to use tragedy to say obsessive no collusion no middle ling , is a prove of is guilty , dear president whit due respect in the world of psychology and psychoanalysis you are lie , otherwise your interaction would have being a good laughter calm an serene and not mention it anymore , now I don't care more than that or at all you are there but please don't go too far to use tragedy and blame the police , and turn around people like this , America in this regard of weapons seems like that it will never change, they have remain at the age of 1800 when they were cow boy , they never evolve they are childish , they like to play with weapons like toy but weapons kills and make tragedy massacre , then they cry out suffer agony anguish , I would like to cry with you but I cant , Americans have gone far away into extroversion into the objectivity, and they have built an ego mind unconscious to huge for any reversibility, I am sorry to say but the collective unconscious of the American is irreversible, no device method can help them anymore they are disease perversion that cannot be heal no matter what device method of meditation you use, sexual perversion for example are irreversible , so an unconscious ego mind that as go too far into extroversion into objectivity and has become to huge is impossible to reverse , the collective unconscious of the American is to huge too big for any reversibility , certain it hurt to see millions of people beautiful intelligent people like the American but so childish that use weapons as toy is childish , and it hurt to see 300 million people like the American in the hand of a few lobby of weapons producer, those producer are just few people powerful of course because they hold allot of money, but they are blind deaf , insensitive whit no inner being or consciousness, that they go on commit crime against humanity for the sake of bunch of dollars is so stupid almost unbelievable because American are rich people why they should care about a bunch of dollars , but any how all over the world the trade of death the trade of weapons is flowering , useless to say every day that is a crime against humanity to build and trade device of death call weapons that kill an extension of your own consciousness living being sacred holy divine , the world should take a completely different direction and make table of peace deactivate mass destruction weapons nuclear weapons chemical weapons atomic and neutron bomb , abolish ban stop , with peace table meeting of all nations and simultaneous deactivate abolish all deadly weapons , simultaneous all nations because actually there is no trust but simultaneous all together could be done , the world nations should change totally the way to interact with each other , today is the law of the jungle that the one who as the deadliest weapons is the stronger , my way that I propose is an interaction between nations through intelligence consciousness awareness peace compassion love basically , the nation should confront measure them self through intelligence consciousness awareness, table of peace deal of peace and deactivation elimination of all weapons simultaneous all together, because the world as come of age as evolve grow in intelligence our DNA as had thousands of convergence of evolution since the ancient time of the primitive barbaric unconscious neurotic men , that tradition of wars military weapons we have to get reed of eliminate, conquering what , you know physic science they have shown us that our planet and solar system is just a small dot infinitesimal into the universe what you want conquer what you want do is ridiculous , come to your sense to all politicians and begin right now the disarmament of the all world before that is to late please , just because I live also in this planet and my community of friends that I belong I would like to see my community of friends to prosper for millions of years , but a new dawn of civilization intelligence meditation consciousness awareness is urgent right now , to build millions of mystery school academy, were one consciousness label less adjectives less contents less for all universe is the only truth direction orientation , we are all animate by a single universal consciousness , fundamental law of the universe intrinsic to the universal body consciousness that is just a pulsation of love intelligence, just a quality a creativity , neutral to gender color race age , no interpretation of the little serious men are possible , consciousness is to infinite to define , I call the little men the ordinary men who live into is mind ego unconscious for this very limited and shrink he can see up to his nose no more than is nose , the ego can see only outwards the ego is the hindrance to look inwards . Anyhow we are running out of time and the world is going the other way around of peace deal and disarmament , I am shouting into a desert I know , but I am also very relax because if a total destruction happen through a third world war I we the enlightened the awakened one into the blink of the eyes and we are resurrected into eternity for eternity to come flowing into awareness into the universal body , body less free of the cage of the imprison of the body, our consciousness of the enlightened one is to huge already to live into a body we just blink the eyes and the annihilation resurrection into the law of eternity is on , what make me sad is the unconscious people that they will have to go for a reincarnation journey, for a rebirth into cosmic unconscious and journey maybe for hundreds of years before to find another womb to reincarnate, that is the failure what I consider my failure the death of the unconscious people and as planet earth will exist no more because a third world war means the vaporization of this planet, the new reincarnation rebirth will happen into other planets, that we know exist no other planet with organic unity life human life, but is mathematics deduction that they are sure say by physic science you will have a new reincarnation in other planets the only way for human being to see other planets and forms of life ha ha , look like that our beautiful green planet earth is the real paradise and it is, you know planet earth is the wonder of wonder of the universe the diamond the only green blue water life place of the universe , so take care why destroy it , that is why I am here still talking , Make no mistake a life lived unconscious will generate an unconscious death they are no short cut in the within reality of inner being and consciousness awareness non being incorporeal body eternity itself , if you have lived unconscious it will generate an unconscious death and reincarnation rebirth , the holy mystic master at the Zodiac say that what you get by your own work you have it , what you have not work out you don't have it , in the within reality it function absolutely like this in the objectivity no you may speculate robing stolen and what you have not work out is yours, but within they are no short cut and this make me immense happy no hypocrisy within your reality will function work , only what you have work hard to have is yours , so make no mistake a life unconscious will generate an unconscious death an consequence a reincarnation a rebirth and that for me is the greatest failure of a life , but you are free to choose an unconscious life for existence freedom is sacred and it respect also your choice to live an unconscious life , existence will go on give you infinite opportunity to be reborn reincarnate infinite body until you got until you will use the opportunity that life is to become conscious aware awake enlightened then is done, no more reincarnation rebirth, then your death will be a conscious death and annihilation resurrection into eternity, a new begin body less into the infinity of eternity and into awareness into the universal body into universal consciousness extend into the objectivity, of who knows which planets because our planets earth you have destroy vaporize with nuclear weapons , but you in essence will have a universal body you will be everywhere and nowhere in particular expand into intrinsic to the universe free, into freedom into an open relativity endless into infinite bliss into infinite silence into all living being into all forms of the universe into all planets galaxies stars, into the wind into the oceans into the nature, animals ,you will be roses you like it in the fragrance of the roses into the color of the roses , that what is the resurrection a conscious alchemy from unconscious to consciousness awareness into non being incorporeal body into eternity itself begin less endless never born or die, and not gross not physical not material , is from here to here from eternity unconscious to eternity totally conscious aware enlightened but basically you are not going anywhere you will be here now flowing into universal consciousness ,,,,,ok I guess is enough maybe I say to much .....Angelo Aulisa

The new book Mysticism & Physics Chapter 48 Angelo Aulisa

Toward civilization part 8

Hi friends , I would like to talk about climate change and echo system , pollution and contamination , this world is in dramatic condition situation , for example India and China they are growing economically at fast higher rate stage , but the price that this two country in particular are paying is an heavy price , there country are totally contaminate polluted not possible to live in India anymore the wild blind economic progress as make all city of India totally contaminate polluted poisonous, in India everything is contaminated poisonous every city as 50 time more contamination poisonous pollution every square cubic meter of air , and to go in India is sure an early death from contamination pollution , I don't think that is worth the price to destroy totally your country for the sake of little money , this is crazy neurotic , also India as a demographic explosion that as reach level incredible , one billion and 350 million this is one of the primary cause of the dramatic situation of contamination pollution poisonous of the country , for me wild procreation as to be abolish absolute stop to birth in country like India , Africa ,China , Asia and middle east , for 30 years the world is collapsing under a clouds of carbon dioxide gas poison of all kind , and recycle of garbage is not happening in any of this country mansion is absolute madness each person produce around 3 tons of garbage every year and produce tons of gas pollution contamination through means of transportation car flight industry factory , we need to rebalance an equilibrium, this planets earth actually can guest one fourth of the actual population, and this country mansion above they go on producing dozens of children every family, if not dozens but 4,5, 6 , blind without thinking of consequence of economic support of the children, is so cruel to make children that after words they will have to suffer starvation no food no close no ruff no toilette living in between the garbage with poisonous air to breath, contaminated air water and ambient totally polluted, me I propose a stop of birth for 30 years , the world need an equilibrium balance urgent right now , and in country like India they live with secular tradition that create such a social difference that few people have the wealth of the all country, I am pointing to the cast system in India one billion and 250 million have no human right, they belong to lower cast they have nothing at all , and the secular traditional religion as create in India a situation of sexual repression horrible , in India they still marriage through arrange marriage they don't know what means love , we the freak that we were living in India we were predicting this sexual repression situation will make India a place of beast and that what as happen now today in India every 25 second happen a case of rape and every minute a case of paedhofilie , horrible a disaster the failure of civilization I am sorry to say but Indians are really unconscious beast , and the country totally a disaster of pollution contamination dirty everywhere , and is irreversible because the population grows at a rate that every 20 years triplicate so India in 20 years there population will be 3 billion and plus , same the population of Africa a and China so the middle east and Asia generally , we need an absolute stop of birth otherwise this country will destroy totally the echo system of the world , they will produce so much pollution and contamination that the climate change will accelerate dozens of time and the ice shield of the planet either north or south will completely melt down rising the oceans water of 350 feet, were many country city will be submerge with oceans water, primary, but they will create a situation where we cannot breath any more , is ok to have progress and evolve but not blindly wild without any criterion is mad crazy enough is enough , stop at birth for 30 years , yes also western country contribute to pollution and contamination but in minor scale much minor scale because the population grow almost none zero growth of population, western country are more intelligent civilize , and the major cause of pollution and contamination believe or not is done by men, and the demographic explosion is the major cause of the deterioration of the gas Serra pollution contamination , . Also the active wars create allot of contamination pollution the bombs that they throw are poisoning polluted the planet the air as nothing else , for not talking of the plastic situation in the oceans the dumping of garbage into the oceans as make the oceans a pound of plastic wastage , there is a big island of plastic floating into the oceans bigger that the America , and it will be there for hundreds of years , and also the deforestation of the world they are cutting trees everywhere the greenery the trees function as filter to clean the air from carbon dioxide , they breath poison and throw out clean air , the deforestation is a big problem they create in balance a disaster , for not talking of nuclear power station that even if they are function properly they create contamination radioactivity, a nuclear power station create the same amount of radioactivity as Chernobyl every 4 or 5 years even if function properly , and every year 7 tons of plutonium and with one and half kilogram of wastage plutonium you can make a nuclear bomb that is the speculation , and the radioactivity of nuclear power station create cancer and skin cancer and DNA mutation deformation and instant death in the worst case and is in the air for you and me to die . Now I question the all world don't you think that we have gone too far and the situation is out of hand, we should rise absolute consciousness about climate change and echo system, systematic destruction done the unconscious barbaric little serious men , the climate change create huge storm that humanity as never witness before and disasters tragedy immense horrible . My propose is a stop of birth right now for 30 years if is not to late already , stop of production of weapons nuclear and conventional right now is a crime against humanity, either production of children, and of weapons too , stop of wars that polluted contaminated the world at large, but with the stop of weapons production also wars will stop as a consequence , stop of military , they can do thousands of other jobs, and also the factory that produce weapons they can do thousands of other item beneficial for humanity like producing equipment for reduce pollution contamination into the world , and the consequence of abolishing completely the production of weapons into the world will give back to humanity 70 per cent of resources of economic resources, try to think no more weapons the resources that were going for death and destruction converted in beneficial item for humanity, we will give those resources to physic and science and they will try the impossible task to restore the echo system only physic and science can do it at this stage , and the abolish of military try to think how much money in salary they get the military a huge mountain of money that they were going for death and destruction converted in beneficial item for humanity we will give all those money to physic and science to restore the air of the planet and the echo system generally, they will build echo biological city at the measure of the new men meditative and mystique , of course we will not waste those resources for launch stupid rocket into the sky into a desert call moon or Mars, that is death since 4 billion and 200 million years there we will send a rover at the most is desert with send storm at 500 kilometer per hour and temperature that in the night fall 100 degree under zero and in the day impossible climate and no air to breath basically , no we will not waste money for absurd mission , the world our planet need to be civilize at large yet ,and that is what we will do with the huge resources that come from the end of weapons production military salary and end of wars that are meant to kill and destruction of our beautiful green planet , the world of the future for 90 per cent will be led by physic and science and 10 per cent by mysticism meditation consciousness awareness , also the production of car of the future will be electric and this is the way I am happy for this, car are the major cause of pollution and contamination I am happy of this go on at high speed all car electric , we will make of this planet the lotus paradise and this body the very consciousness awareness, and breath clean air , and live 140 years ha ha ..But not so much, the situation of echo system is dramatic with urgency we need a total transformation right now. Me from my side I eliminate completely now and in the future to go in country like India I am sorry for Poona the community, but Poona is one of the city of the world most polluted on earth is totally not healthy to go there and live there for me is over and over I have love the community life but is located on hell and with due respect who want to go to hell wen paradise are available.....India is over for me and all country were life as become impossible due high level of contamination and pollution India is like a poisonous snake all contaminated and dirty sorry but the truth is the truth, one friend of my as tell me that is friend as gone in India in Mombay and she was totally shocked at the contamination pollution dirty i say is not my responsibility but that is the crude reality bad smell dirty pollution contaminated everything food air water, even the water they sale on bottle is horrible so i will not come ever again for me is over who want live in a place were every 25 second happen a rape case and every minute a case paedhohil the failure of civilization i refuse to go in a place of unconscious beast sorry you can have all for you all forever but not me ever again ....Angelo Aulisa

Published on February 21, 2018 05:41 • 15 views

The new book Mysticism & Physics Chapter 49 Angelo Aulisa

Toward civilization part 9

Hi friends , we continue our journey towards civilization , and so many country of the world and city of the world have become place were life is not anymore possible to live due the immense contamination pollution poison, of every square cubic meter of air an average of 50 time more than the normal average allowed , I have make a small research on computer to check out the figure picture of contamination pollution and is dramatic almost all over the world, for example place like Turkey all city are totally contaminated polluted poisonous , Europe is still livable not everywhere but many country are livable , the dramatic figure come out in Africa China India middle east Asia and south America Mexico my gosh and Russia , in short almost all the world is unlivable , of course out of the city in mountains the situation change a little , is better to be conscious of such epidemic of climate change and pollution contamination to choose the best healthy place to go and to live , and to spread rising of consciousness for the urgency all over the world to change do something drastically to transform this horrible situation condition of the world , also because the consequence are going to be dramatic , melt down of ice shield north and south pole, rising of the oceans level of 350 feet , disappear of many city and country submerge by oceans water , deaths of millions for respiratory disease like cancer etc... huge storms as never witness before , radioactivity from nuclear power station that will never go away the nuclear radioactivity once create it stay for thousands of years, many horrible disease are the consequence , the dump of toxic score everywhere and most into the oceans, and the dump of plastic everywhere and most into the oceans the one who eat fish he eat really poison already , and consequence disease , the deforestation of the planet create toxic air hundreds of time more because the trees are filter of air they breathe carbon dioxide and throw out clean air , the disappear of glacial due the rising of temperature I love the glacial , and the wars that they do not only create deaths and destruction bloodshed but also immense contamination pollution , the weapons are death poisonous device a crime against humanity to build them , so the situation is dramatic maybe irreversible if the humanity don't act right now urgent to prevent such a disaster tragedy of the world , I still write but the direction orientation of the world today is towards self-destruction this so call humanity will self-destroy itself , the within insight reason is because men humanity live in anguish frustration agony, totally unconscious into the ego mind unconscious and life as become almost for all the world a tragedy a suffering a misery, due because men humanity don't know how to answer the basic question of the inner reality of men like , who I am ,were I come from, were I am going ,what about death the fear the anxiety agony that death create in people is immense the restless that death create is kind of panic restless , and until you come to terms with death you cannot live a peaceful tranquil life, you will be restless in agony anguish frustration , and more were I am living because most of people don't know anything about the organism of the universe and they don't care or they think that there is no relation between human being and universe or planet earth itself, they have no link no bridge that is what I am trying to create make almost everyday bridge link with the organism of the universe and the organic unity that a human being is , without this roots into your inner being and consciousness and non being incorporeal body and awareness into the law of eternity core and source of the mystery of the universe and of life and death and of all duality , without this roots into meditation silence love , men is like a trees without roots throw there to the sun to dry up and die , and the agony anguish frustration misery is huge immense, so men with a saturate individual unconscious and a saturate collective unconscious that is stem up, is totally lust in a mess confusion chaos and the pain is unbearable , so is ready for a self-destruction who want live like a tree without roots throw to the sun to die , and who want live a life that is a story tell by an idiot full of fury noise signify nothing at all , right Sartre to say in is philosophy of meaningless he say life as no significance meaning, as it is today is right a life lived only on the surface of objectivity material only on the dimension of having eat drink and be marry , only in the horizontal dimension of ego mind unconscious at infinite regress linear horizontal as no meaning significance at all, hence the frustration agony anguish , George was ask where is the point he answer there is no point at all , because only objective love for life is not enough , love can help you to expand inquire to other dimension of life, like the vertical dimension of deeper and higher of here now of the moment of the present gate less gate to your inner reality to your inner being, you close your eyes and bring the attention within the so called dark matter or gravitational energy field will bent the light consciousness say Einstein deep within your inner being, were you will realize that you are not your body not your mind not your thoughts not your emotion not your unconscious staff not your senses, but you are just a crystal clear witness consciousness, an empty mirror that reflect all and everything that surround you without charge or comment, just a crystal clear witness consciousness link bridge from the known to the unknown to the unknowable, and deeper into your inner being into pin drop silence, is the gate less gate to your inner universal consciousness, were your inner being annihilate into universal consciousness and you are the universal body the intrinsic consciousness it ,the so called dark energy or expansive energy field will expand your light consciousness higher and higher up to the horizon event of the universe and your size now is the universe itself in essence you are the universal body the universal consciousness, and the so called dark flux will flow move circulate your light consciousness in a cosmic play overlapping with non being body, the universal body that you are now in essence breath into non being body incorporeal and non being body breath into the universal body and the light consciousness that you are , the dark flux trigger this majestic cosmic play is divine sacred holy, then emptiness nothingness are the gate less gate to non being body were forms time space duality totally annihilate, and consciousness too annihilate consciousness is always in relation to a subject or object no more subject or object into non being body and consciousness turn twisted into unfocused awareness, relation less just an I am ness, everywhere and nowhere in particular expand into non being body ultimate essence of it, then you are at home into infinite freedom infinite bliss infinite peace into an endless open relativity, sacred holy divine into infinite silence intrinsic with subtle ecstasy , the size non being body vanish into infinity ultimate equation of the quantum mathematics, and that is your size now in essence infinity infinity were you will realize happen immortality resurrection , that what your DNA aspire for the ultimate convergence of evolution of your DNA, and also ultimate aspiration of the pulsation of the light consciousness and this is the triumph of civilization a man as become conscious aware immortal resurrected enlightened forever and ever , you have accomplish the aspiration of existence of your light consciousness you can have a cup of tea now and celebrate ha ha , I say all of this because life is much more then material objectivity, life is an endless journey that end no where, into eternity, life is a transcendence into eternity itself a great mystery to love to celebrate to delight in a sacred holy play , and until you will have this inner experience journey through meditation or love or silence the anguish agony frustration will continue the misery suffering will continue and there is no need to destroy the world to self-destroy yourself and the world , we the enlightened one awakened one we are here to shier light for you all, is almost a race between the old little serious unconscious blind deaf insensitive man who live in is unconscious ego mind, and the enlightened one who shier light , to the world who shier consciousness awareness to the world we the enlightened one we have no power objective of any kind, the unconscious men as all power material but is blind let do like this the house is on fire, the wife is blind and the husband is lame, the house is on fire burning so the husband say to the wife you take me on your back and I will show you the way to run out of the house and so they do it and the safe their life ......Angelo Aulisa

The new book Mysticism & Physics Chapter 50 Angelo Aulisa

Toward civilization part 10

Hi friends , we continue our journey towards civilization , so the world as become unlivable due the climate change consequence of pollution contamination poisonous done by men , now I want say that there is out wards pollution contamination poisonous of air ambient echo system generally , and there is inwards pollution contamination poisonous create by system society ,, out of date expire superstition of religions that are totally insane lies and lies and neurosis, of expire out of date dead tradition , with conditioning repression wounds trauma lies and lies of all kind, false information false notion, in short with a bunch of neurosis , they feed your neurons they projected in your neurons all of this staff , they neurons function as vector of thoughts emotions believe in a chain reaction into your brain, and once done this projection of unconscious staff your brain is damage forever this projection in your neurons constitute make your unconscious, that accumulate new dust every second on and on . Now I want report a news that one month ago was give on BBC , in India a baby of 8 months was brutally rape do you believe , sorry for the disgusting feeling horrible horrify , but this in India is only the top of the iceberg of an epidemic disease which affect the all population of India , the sexual repression create by their own secular false ugly bigot fantasy out of date expire religion call Indus , India is not only the most polluted contaminated poisonous place of the earth out words into the objectivity but also within the inner reality , they have a multi theistic religions that are all fairy tale for retarded children , they have many divinity the trees the animals the rocks etc... but is all an hypocrisy because when it comes to human being the caste system put the lower caste as slave with no human right ,with not thing at all as having into the objectivity, in India they are one billion and 300 million people treat as slave no human right, and most of all the higher cast of stupid idiot ,they cannot even touch a lower caste from Paria, not even the shadow if the shadow touch one higher caste they have to take a shower , they are not divinity, a monkey is a divinity but not a human being great contradiction hypocrisy , and they use them for the most dirty work of the society , with such a violence and rage that you cannot believe I have live in India 30 years I can tell that is like this , and for not talking how they treat woman factory to reproduce children , and there tradition bigot ugly out of date expire for example of Sati the woman who get marriage if the husband die she as to jump into the funeral pyre into the fire when they burn the husband alive , but not vice versa if the wife die , they still sacrifice human being to divinity in part of India and sacrifice animals to their bogus divinity ,the pretension of the Indian to be the most spiritual country is a laughing stock worldwide they have no consciousness no inner being they are simple unconscious beast dirty , in India every 30 second happen a rape case and every minute a paedhophily act , due the their incredible sexual repression , the Indians see a girl and touch a girl only through arrange marriage at the age of 30 , 26 , 27 , before that they have no any kind of experience , naturally it create a hell of repression sexually , they consider the foreign woman all bitch unconditionally and a foreign woman who goes in India is on is path to hell to be rape assault brutally in gang rape , and this make of India the most fascist country of the world, the cast system , what they say that India is the greatest democracy of the world is the biggest lie ever utter, India is the most fascist country of the world Nazis I would say , So in India there not any kind of spirituality all lies and lies and fairy tale for retarded children , and democracy is dirty word it do not exist , what exist is a fascist system with a mask of spirituality so ridiculous , In India they reject completely the approach of the Buddha they cannot conceive in fact Buddhism as flower in many other country but not in India . My beloved master Osho was not an Indian in first place, and he was attempt for murdered different time , when he goes around India to talk they don't let him go down of the train they stoned the train, and he was trying to help the lower cast but soon he realize that was impossible to help them because this hypnosis of the cast system as go so deep that is irreversible , they say if you accept your Karma of being born lower cast you will reborn on a higher cast, bullshit to dominate the people, but the lower cast being illiterate they believe blindly , so my beloved master stop to help them due the dangerous situation that this create , Friends don't gate deceive India is the most materialist country of the world were count only money, and their attitude of the Indians towards the foreign is to cheat wen ever they can , say by experience , I talk Indus language for 60 per cent I meant what I say , in India there nothing spiritually only a big mask that fortunate is fading away more and more , Sorry but the Indians are the most unconscious beast of the planet and there country incredible polluted dirty contaminated due there greedy for money Indians are very greedy materialistic people only that the truth towards civilization .....Angelo Aulisa

Published on February 23, 2018 02:04 • 15 views

The new book Mysticism & Physics Chapter 51 Angelo Aulisa

Towards civilization part 11

Hi fiends , we continue our journey towards civilization , also because sometime amazon.com sometime pay me little royalty nothing just few coins but I appreciate not for the money are just two coins but because give courage support at least 5 people may read my books , thank you , even if I know well that non official I am the number one in the world today, as far writing is concern on my subject of mysticism and physic, also because I have given to the world of physic new intrinsic law as the law of dialectic a master key for physic to unlock the greatest mystery, starting from the big bang before the big bang and after the big bang the law of dialectic of opposite complementary is a master key for the universal body harmony, and explanation ,and many other insight I have contribute to physic , on mysticism I am an absolute emperor programmed to be an emperor , and so in the begin it was void cold frozen void I remember you that the temperature of the space of the. universal body is mine 270 degree and beautiful in the universal space weapons don't function it should be so also into planet earth I wish so then so many idiot imbecilely they will have to find other jobs work, and planet earth would have being a paradise, and absolute emptiness , ... nothingness that at the apex of infinitesimal emptiness it create a singularity small then an atom, that is the primordial thesis of the universal body that at the apex of infinitesimal create is complementary opposite of fullness wholeness antithesis friction of those micro ,macro, and synthesis of big bang, then explosion which unfold simultaneous all the law intrinsic to the universal body inflection that is on even today after 14 billion years, and the begin of time and space we measure time and space from the big bang on ward before the big bang time less space less no forms no duality, the dialectic of duality begin after the big bang , and then dark energy expansive energy field, and then dark matter gravitational energy field, which create a perfect complement with dark energy, the friction of thesis dark energy and antithesis dark matter synthesis the harmony balance of the universal body, and I suggest that this complementary of dark matter or gravitational field and dark energy or expansive energy field friction of those and synthesis of harmony balance and Boson x the quality that hold keep the whole universe together and give Mass to particles and atoms it unfold always simultaneous to all other intrinsic law of the universal body, without the Boson the universe and all of is forms disintegrate , it cannot be otherwise because the big bang generate a clouds of gas dust debris and this cloud is keep holding together by the Boson the particle atoms of the dust gas debris is create by the Boson x , that in mystical terms is the intrinsic inner consciousness of the universe the fabric of life, whole is consciousness remember and all is holding keep together by the boson x the inner quality creativity pulsation of the universe you can call the Boson inner consciousness and you can call the inner consciousness Boson x the same quality creativity pulsation of intelligence love sacred holy divine , and simultaneous unfold the dark flux or flowing energy field that is a cosmic play of the universe the inner flow the inner movement the inner circulation stream of the universal body , and of course the law of dialectic fundamental law of the universe the inner nature of the universal body, all of this intrinsic law unfold simultaneous at the big bang and it unfold the sleeping age of the universe or dark age for two billions years the universe was just in a phase of assembling itself the dark matter or gravitational energy field play the major role in this sleeping dark age to assemble the first stars planets, and consequence galaxies of the universe the dark matter is the greatest painter of the universe it took two billions years to see the first formation of the early universe and further on all the unfold of the universe that you see today, it took billions of years to come together to put together the majestic beautiful picture of the universe that you see today , all the above mansion intrinsic law of the universal body play there role in the formation of the miracle call universe . But as physicians say the sharing of the universe consist of 24 per cent is dark matter or gravitational energy field and 70 per cent is dark energy or expansive energy field and 6 per cent is the matter that you see just a pale shadow of the intrinsic reality of the universe, and this sharing is not balance , the universe first of all is a great play of opposite complementary say Einstein and Manik and the law of dialectic of Hegelian opposite complementary confirm it , So we get 70 per cent dark energy expansive energy field that let say is pulling upwards is telling apart the universe thesis and 24 per cent of dark matter or gravitational energy field that is pulling deeper inwards antithesis friction of those and a balance harmony the Boson x the inner consciousness is the synthesis that unfold the harmony of the universe I remember you that harmony means................ cosmos literally , and this synthesis is what keep the universal body together in harmony balance, the boson x and inner consciousness come to be in relation of the subject of the universal body otherwise are dispose into non being body incorporeal body, into the ultimate canvas of the universe the law of eternity were actually the universe id display, and eternity is space less time less forms less duality less and core and source of the mystery of the universe and of life and death and of all duality, just the ultimate equation of the quantum mathematics infinity infinity, an endless freedom, bliss peace silence, sacred holy divine an endless open relativity not absolute at all because it begin or end nowhere , but with the event of the big bang the Boson x and inner consciousness come to be simultaneous to all the other intrinsic law of the universe in relation of the subject of the universe they unfold because they give Mass to particles and atoms no boson x no consciousness no particles no atoms, are the quality that give Mass to matter , what the little Kuku idiot unconscious ancient men call God , and this friction of dark matter thesis and dark energy antithesis and friction an synthesis of the universal body harmony balance, cannot go on for so long say the physic scientist the difference of share is to big the 70 per cent of dark energy in the long run will win the race and will tell apart the universe , in 60 billion years let say the universal body will turn again in a cold frozen void, if you were there you look up at the sky and just darkness cold frozen void emptiness nothingness as the primordial thesis before the big bang , this momentum will last billions of years until wen the void emptiness nothingness reach is apex of infinitesimal, micro, were it create a singularity smaller than an atoms, infinitesimal were will create again is complementary opposite of whole fullness macro, the antithesis and friction of those and again a bang a big bang and a synthesis of a new universe come to be, and again all the cycle of all the law of the universe unfold , is an eternal journey that end nowhere also for the universe due the fundamental law of dialectic of opposite complementary Hegelian which is the fundamental law intrinsic to the universal body brought to light by Manik the great , and give to the world as master key to unlock the greatest mystery of the universe , as for consciousness awareness death do not exist is an eternal journey cycle that begin or end nowhere so is for the universe in large scale a never ending pulsation of intelligence love creativity , also the universe is immortal resurrected ha ha eternally flowing never born or die never begin or end , many universe have already being and many universe will be again and again, each universe that come to be is like a season unfortunate this universe our universe is the season of the unconscious idiot little imbecilely men serious bigot expire out of date you like ha ha, we need a urgent new dawn of consciousness awareness civilization intelligence meditation, we need to reclaim the evolution of our DNA and consciousness that as take thousands of years to evolve to refine to mutate, we reclaim our intelligence evolution obtain with hard work and sacrifice , we need to dispose of all this idiot tyranny dictators that block the evolution of humanity they hinder the evolution of DNA and consciousness, with their ignorance there unconscious they create the age of the dark unconscious enough is enough resign all of you dictator tyranny Putin ,Assad, Erdogan, the dictators club in generally they are dozens around the world right Bill Clinton there is a club of tyranny dictators that hinder the evolution of humanity in intelligence consciousness awareness, and of our collective DNA generally, for the good of human Kind resign live and let live, please resign right now all of you the game is over you are commit countless crime against humanity that you will be account of because they are show in digital color simultaneous every day in TV all over the world, who do you think to turn around stop genocide right now , Anyhow they have being other season of other universe, but none of them as reach so higher as our season of this universe we live in today, we have the chance to create the highest possible civilization that as ever happen in any universe that as being, the diamond age of oneness mystical union into eternity itself men in essence is eternity itself immortal resurrected enlightened forever and ever ha ha ...Angelo Aulisa

Published on February 24, 2018 02:20 • 13 views

The new book mysticism & Physics Chapter 52 Angelo Aulisa

Towards civilization part 12

Hi friends , we continue our journey towards civilization in spite of the genocide that is gone on in this day , in Syria perpetuate by Russian Putin and Assad under the eyes of the all world , And so gravity bend light and time and space too creating gravitational waves all around like throwing a stone into a lake that create ripples all around , Einstein pointed at it many years ago , and 2 years ago I declare it to the world on 2,02, 2016 , introduction of Les siècle des lumieres my book , 14 days after the BBC give the news of declaration that somebody have discover the same thing copy from me , and last year it win the Nobel prize for science but is not the first time that happen many Nobel price the inspiration as being take from me Manik the great ,. I am not hungry at it even Einstein when he discover the theory of relativity they was only twelve person all over the world that understand it , anyhow he was ask how you come to this discover and what means the theory of relativity because actually we don't know , Einstein is reported that he answer like this , first of all if I was not the one who discover the theory of relativity within twelve day someone else would have discover it , and in this answer already is enclose the meaning of the theory of relativity but he continue and say that the theory of relativity is like this when you are with your beloved loved friend one hour is like one second and when you are seat on a hot stove one second is like one hour do you get or got German ……the post that I am going to right this evening is about past life because to me in a journey towards civilization is important to include the truth of the journey beyond death true and accurate , So we get a human being that build an unconscious individual from the first day that he born onward natural consequence of the feeding projection into is neurons, of system society family, out of date false superstition prejudice religions , conditioning repression false notion false information, wounds of tradition expire false trauma of tradition religion false expire out of date , out sources generally , the neurons function as vectors of thoughts believe emotions in a chain reaction into your brain and this content projected on you constitute the foundation of your individual unconscious, once done your brain will be damage forever, and the unconscious goes on accumulate dust every second on and on for all of your life and is a huge iceberg you see only the tip of it one tenth, nine tenth are under net of it you don't see it but is there like an iceberg . Then we get or got the collective unconscious, they are dozens of collective unconscious into the world every nations has its own collective unconscious, every expire out of date religions as is collective unconscious, different situation circumstance different collective unconscious it enclose on it all the tradition dead of the past, all doctrine philosophy of your expire religion all tradition of your society system, all tradition of your family of all the past lived up to your day to day life , it give behavior image character attitude to life, once you are condition by it you cannot escape the collective unconscious is huge immense if the individual unconscious is an iceberg the collective unconscious is the all continent where you live ,once you are take condition by It you can say good bye to freedom naturalness spontaneity originality forever ,and ever , this is the dimension of the known humanity knows about individual and collective unconscious, Freud Carl Gustav Jung are the fathers of it , the discovers of it almost since one century and are very accurate analysis were you can bring on the surface of light of life society all the wounds trauma repression conditioning of this two reality of unconscious , I myself at priory I consider always the unconscious of the person who interacting with me, I don't denied it never rather I study research each single case and discover many aspect hide to the eyes of men, of the person who is interacting with me , a tragedy a disaster a bunch of horrible neurosis all the time surface and create an havoc on me, but that is the way to protect myself sometime is dangerous because digging deep into the various unconscious you discover that the person who is interacting with you is a monster or psycho and many other dangerous disease so you have to be careful , but I never denied the reality of the unconscious and if you denied you are in for a disaster or horrible circumstance situation sometime very dangerous take care . Life is a big laboratory of analysis and psychoanalysis. Then let go further into the unknown what humanity do not know or record nice , the cosmic unconscious or spiritual body I call it conscious body , that goes in an inner journey in retrospective up to the super conscious or non being body , it consist of many layer , the first layer is of your past life thousands are enclosed into the cosmic unconscious, and all is record in a factual memory in digital color , when you die the death create a cur tine a filter with the previsions life out of necessity so you don't remember the previsions past life , otherwise you would go crazy because is already an heavy load to remember this life unconscious try to think if you remember all of your past life you will go mad , but they are method to pierce the cosmic unconscious meditation if you are sensitive alert intuitive and vegetarian, or dreaming if you are sensitive alert awake as witness consciousness into the dream , but if you are awake as a witness consciousness into the dream the dreams stop ceases, or hypnosis done by therapist , or when you make an accident and you fall into a coma if your witness consciousness remain awake you can witness past life conscious but this rare happen in a coma actually you fall unconscious , or Dejavus moment, are moment that you relieve past life you was there before you remember you have a flash of remembrance of your past life , anyhow you can journey into your past life into the cosmic unconscious and relieve many of your past life if your witness consciousness remain awake you will remember when you come back into the body, and this will give you a certainty that death do not exist if I was before in different forms body that means that my consciousness is a continuity of many life so I was I am and I will be , this give great relaxation peace that death do not exist and this is the mean point of having the experience a freshness a new begin that life is eternal , and second you can unfold many wounds and see the relation with your present life for example you may have being a prostitute in your past life or a thief or a murder and so on so forth so this give clarity about so many block that you have in your present life to clear many wounds of your present life is an ha experience ha that is why , I am so and so , and more if in this life you are a man in your previous life you were a woman and so on so forth for a law of dialectic it unfold like this generally but not as a law they are exception , more the way to record nice that you are journey in past life is that the journey unfold back words and not forwards from the last life that you have backwards on and on this is a great mile stone to see super conscious or non being body 3000 miles , you are on the right path , Then when the layer of past life exhaust finish you jump into the animals layer because you have being many kind of animals too if you pierce deep and remain conscious you can see clear what animals you were a tiger a dog a pig I think many have being pig ha ha joking then always backwards the layer of nature trees plant flowers basically at this stage become a vortex of colors, then the layer of rocks consciousness is present into rocks asleep but neither the less present unconscious asleep in evolution , then the layer of water consciousness is flowing asleep into water flowing unconscious but present, then fish because fish come in the end because fish is the first forms of organic unity that take your unconscious energy journeying towards refinement of other forms in the water in the oceans or lakes, then the layer of bacteria microorganism amino acid, then the irrational jump quantum leap into super conscious or non being formless, no forms just an I am ness just a relation less awareness unfocused then you are everywhere and nowhere in particular diffuse annihilate into the super conscious or non being body incorporeal or call it eternity begin less endless never born or die , and the journey is over you are at home into the core and source of the mystery of the universe and of life and death and of all duality in infinite bliss freedom peace ecstasy silence sacred divine , diffuse into an open relativity in the within of the universal body and all of is forms , and if you still hold a body at this stage and the body is healthy, and if somebody recall your light consciousness you may return into your body but maybe even not because the bliss ecstasy of being infinite is so much that you don't care to return into the body unless someone call you back lovely touch your body lovely , and if you don't hold a body because death as occur so far so good you are resurrected for eternity to come you will flow intrinsic to eternity diffuse annihilate everywhere and nowhere in particular into infinity into eternity into awareness into the universal body into consciousness which extend into the objectivity of the universe into stars galaxies planets wind flower living being all forms of the universe generally intrinsic within flowing through consciousness body less of course, in data intelligence, quality fragrance that you have refine in your life that the resurrection a conscious alchemy not physical not gross not material, and if you go the cemetery you will find all the body in putrefaction eat by rats and worms without exception , stinking because the soul as migrate away , or if you hold a body and return into it you will be enlightened forever and ever you are resurrected into the body you will live a life of bliss freedom peace silence lovely pulsating in love intelligence the circle is complete the zero is complete you have return back home totally conscious refine a light consciousness, life become a sacred holy play divine ecstatic nothing anymore is serious but just a playfulness sacred divine , One thing very important if you have reach the form of human being the reincarnation will be always and always as a human being there is no regression into other layer of the cosmic unconscious, for a natural law of evolution as is our DNA always evolving mutating in new shape in accordance with the new situation that he face forward and forward aspiring for the ultimate convergence of evolution immortality resurrection ,same as the pulsation of consciousness aspire always for the ultimate refinement into the law of eternity into immortality resurrection the journey is from unconscious to consciousness to awareness into the law of eternity from here to here from eternity to eternity but basically you are not going anywhere you will be here into the universal body, body less formless . The reincarnation happen if you have lived an unconscious life it generate an unconscious death and consequence the rebirth reincarnation , but if by hard work you have become conscious aware and you have lived a conscious life it will generate a conscious death and consequence resurrection formless of course , into eternity begin less endless never born or die , if reincarnation take place it depend how loaded is your unconscious actually a loaded unconscious today days find soon a new womb because humanity is living at lower body and a loaded unconscious need a lower body to reincarnate like physical body astral body or mental body etheric body, the world 99 per cent live at this profundity so you may find soon a new womb , and the contradiction is if your unconscious is not so loaded you may need longer time to find a body a womb because in this humanity today days nobody lives at higher body like spiritual cosmic, or not non being because non being you would not reincarnate, so a less loaded unconscious need more time to find a new womb and loaded unconscious very soon find a new situation because is heavy and lower and so live this humanity today, you will reincarnate always from where you leave your body, from lower body then lower idiot who make love sex at lower stage, if you leave at higher stage the body then you will reincarnate with a situation of higher body you need a situation of lover who make love at higher body like spiritual of cosmic, not non being body it do not generate reincarnation , what you get by your own work is yours forever and ever you will not lost what you attain in the present life even through death it will be yours forever and ever, that why you need a couple who is at an higher level of body as spiritual or ....cosmic and is not easy as the lower body death and reincarnation this is a great contradiction but it is so a refine soul will need longer time to reincarnate and an heavy soul will reincarnate soon ......Angelo Aulisa

The new book Mysticism & Physics Chapter 53 Angelo Aulisa

Towards civilization part 13

Hi friends , we continue our journey toward civilization , and past life , I am going to give some evidence for me I don't need any evidence because I am talking by my own experience I am not quoting theory , one of the greatest evidence of past life exist is the love act , when a couple make love in the moment of orgasm the orgasm the sperm of man contain hundreds of thousands of living cells that in a rush they go towards the egg of the woman , in the race rush only one cells reach the egg and the other hundreds of thousands of cells souls they annihilate again into the cosmic unconscious , now who do you think they are those hundreds of thousands of cells souls that do not reach the target of the egg , they are millions of souls that are racing rushing again and again to reincarnate but they miss the target of the egg they are cells souls that are floating into the cosmic unconscious waiting for their own turn to return into play into the objectivity of life , they are all of them potential cells souls that aspire rush race for a new rebirth reincarnation , but only one reach the target of the egg of the woman , a joke , in the infinite space of the cosmic unconscious they were are many souls floating around and journeying around waiting for their turn to return into play into life on planet earth and it was Peter soul that was doing everyday exercise training running and it was George soul that was watching the all story George soul was very lazy always relaxing under a cloud, but was watching Peter soul running training so one day get courage and George soul ask Peter soul sorry but why you are always running training and Peter soul answer because wen is going to come the time to run for rebirth I am going to be the first, as I am training to run every day, there and then the bell ring and all the souls of the infinite cosmic unconscious start to run towards the target the egg , and Peter soul of course disappear rushing racing and George soul was always lazy at the end the last of the racer , , after a while he see Peter soul coming back with a long face disappoint sad and so George soul ask what happen at which Peter soul reply ha fuck it was only a blow job . Any how the millions of soul's cells that miss the target of the egg is a clear evidence of the cycle of reincarnation rebirth. As I always say a life lived unconscious they are no short cut it will generate reincarnation , but a life lived conscious aware meditative can be a resurrection into eternity is not a law they are exception , few secret how to enter death conscious may help , one of the most important thing is that at the moment of death don't resist death if you resist death is because in you there is still desire to live life and if on you there is still desire to live life as 99.9 per cent of case and you resist death you will generate reincarnation rebirth because existence respect totally your desire your freedom even a small hint of desire of remain a little longer into life will generate a new reincarnation, even if you have lived a conscious aware life , at the moment of death collect your light consciousness within your inner being and be centered relax and let go into your inner being into the deepest point of your inner being, a state of desire less is require if you have lived a life conscious aware it will be there a state of desire less, that means that you are ready to let go and be conscious let go into the deepest core of your inner being into silence pin drop silence, if you have lived a life conscious meditative aware the desire in you will be empty just the will of life will be there object less empty, and you are ready willing to let go easy and without hesitation let go into the deepest core of your inner being into pin drop silence and be centered within , your inner being is the bridge link to the all existence when you are centered into your inner being you are centered simultaneous into the all existence that mean the all universal body and non being incorporeal body, and the law of eternity itself core and source of the mystery of the universe and of life and death and of all duality , , so be centered deep within your inner being and silence gate less gate to your inner universal consciousness, your inner being will annihilate into universal consciousness and then nothingness emptiness gate less gate to non being body incorporeal ,were time space forms duality totally annihilate disappear then you are at home into awareness unfocused relation less a great relaxation an I am ness into eternity begin less endless, you in essence are at home into the core and source of the mystery of the universal body, into freedom into peace bliss into infinite silence into infinity, resurrected your essence data quality fragrance of a many many life diffuse annihilate ultimate and finally into the core and source of eternity, in essence you are back home for eternity to come this a new begin formless into eternity, into awareness into the universal body into intrinsic to universal consciousness you will flow eternally into all forms of the universal body intrinsic to it to all forms , resurrected formless in total freedom expand everywhere and nowhere in particular into the universal body and eternity , a new fresh begin formless you in essence will have a universal body . But the point that I was going to point out is that when you are centered deep into your inner being and deep into pin drop silence you will hear within you a soundless sound of AUM, the ultimate vibration of the universal body let go easy willing welcoming into the soundless sound of AUM if you do that easy welcoming willing in a let go the inner journey into eternity will unfold simultaneous and easy, like when you fall into a vortex of water into a river if you resist you will drown die if you let go into the vortex of water into the river like as if you are death it will take you down at the deepest of the vortex but then suddenly it will reject you upward into the surface of the river and u are save safe , let go easy desire less welcoming willing into the soundless sound of AUM and simultaneous the journey into eternity and resurrection unfold natural , this skill you apply into life into the thousands session of meditation that you go through, meditation done for real is death experience and if you are training for thousands of session for the experience wen real death actually happen for you will be just the blink of the eyes easy you let go, remember let go is the secret of meditation if you have skill neck of let go for you death will be the greatest experience of your life a liberation from the cage prison of the body your consciousness finally goes for the ultimate journey into eternity and resurrection , a celebration death is a great celebration for an enlightened light consciousness finally it let go into eternity and freedom real freedom into bliss sacred divine into peace into silence into infinity and finally formless into infinity, it will be in essence data intelligence fragrance for eternity to come, the resurrection as happen gone through for eternity to come you will be of course formless as an essence fragrance quality data intelligence, intrinsic to the universal body and eternity itself and awareness, into universal consciousness into galaxies planets stars and all forms and living being of the universe ......

Angelo Aulisa

The new book Mysticism & Physics Chapter 54 Angelo Aulisa

Towards civilization part 14

Hi friends we continue our journey towards civilization , and civilization the priority of civilization is a world without weapons of any kind no wars no weapons no military is the first step towards civilization no first step no civilization , the politicians should must stop the attitude of been broker dealers of weapons and consequence graveyard diggers they have dig a big graveyard for the all world with their death trade of deadly device call weapons weak up stop digging grave ......make peace table deal of all nations together for disarmament of the world all together simultaneous we Can because basically there is no trust but all together simultaneous we Can .....So the post of today how when you enter death conscious a desire less is require when you are desire less what is left is the will of life, empty of object or subject and empty of emotion and emotional attachment like parental feeling of emotion towards sister brother mother father beloved loved friends etc... you born alone and you enter death conscious alone, that is a basic require wen the desire is empty completely you can call also will of life or empty consciousness awareness also apply, and let go in the within core of your inner being center also simultaneous of the all existence that means your inner being consciousness awareness non being incorporeal body, were forms duality time space annihilate disappear, if you from a stage of inner being let go trustful willing welcoming easy natural the inner journey towards resurrection unfold happen ....Anyhow I was going to write about mysticism without physic is lame shrink, today days because physic in the last twenty years as evolve expand immense into give new revelation discover to the world ,like the Boson x that took 40 years of hard work sacrifice of thousands of scientist physicians and hundreds of millions of dollars of investment but finally 3 years ago the experiment went through happen , and physic as given to the world the revelation discovery of all intrinsic law of the universe like dark matter or gravitational energy field or dark energy or expansive energy field or dark flux or flowing energy field or the inflection that is expanding the universe even today and on , or the orbit the different orbit of planets stars into the universe very important to define study them , or the magnetism of the planets the magnetic field of planets stars very important to define it and so on so forth the fundamental dialectic law of the universe, physic science as evolve more in the last 15 , 20 years that in the previous 350 years before, since Galileo that he brought many new discovery most of all that planet earth is a round planet and non-flat , the general standard theory of relativity of Einstein, and further the string theory of all and everything of the new physic , a great immense evolution thank you very much to all physicians and scientist your contribution to the world is ,immense , what is important to understand that all of this intrinsic law of the universe influence and determinate the organic unity of human being and life , all of this new discovery are not something apart for human being but they constitute a new living in a new quality and understand of human being, the organic unity of a human being is link connected bridge intrinsic to the organism of the universal body they unfold in a mystical union oneness , and they give a new understand vision of life to humanity, and the organic unity that a human being his , for example The Boson x hold keep together your body too all the particles and atoms of your body the Mass of those particles and atoms are given by the Boson x and they hold keep them together same as the quality of the Boson x hold and keep together the all universal body and confer Mass to particles and atoms not only of all the forms of the universe but your body is included on it, no Boson x an all and everything the universe disintegrate instant your body included , The boson x confer Mass to all particles atoms of the universal body your organic unity included , your body is a play of particles atoms compounds , so the gravitational field call dark matter influence and link connected bridge your organic unity first of all it keep it assembled together one, and further it link your organic unity to all universe is a gravitational field expand into the universe everywhere and nowhere in particular it constitute the 24 per cent of the universe, and it pull inwards deeper within it bent even light , consciousness is light actually light consciousness, and the gravitational field bend it affect it and so those the expansive energy field call dark energy, that is 70 per cent of the universal body, it pull upwards higher and higher your light consciousness and create a friction synthesis with the dark matter an harmony , balance , , so those the dark flux the flowing energy field of the universe is intrinsic to the universe a cosmic play irrational, of movement flowing circulate your light consciousness actually it breath in and out of non being body your light empty consciousness ..... and so on so forth what I want point out is that physic and science have mutate change completely the quality understand of living of human being and they have create a great quantum leap ,mutation convergence of evolution in the life of all human being organic unity forever and ever, we cannot deny this great convergence of evolution that physic as brought up just to make survive the out of date expire religions superstition prejudice religions, they have done there time they have expire totally demystify by intelligence of physic and science that by the way the

religions have been always antagonistic an hindrance a block for physic and science since Galileo on wards that they kill Galileo has been torture and put in prison by the Christians religion, because in the bible was write that the world is flat and if there is one lie in the holy book that put the whole book on doubt of lies, and that how it is the bible is all lies and lies and since then the all religions have being against physic and science , but today days I ask the resignation of all religions unconditional because they have been demystify, expire out of date by the immense evolution of physic and science , and so today days physic is meta physic , and as come together with mysticism, but physic is blind without mysticism one is lame mysticism without physic and physic without mysticism is blind , same tale of the house on fire the husband is lame and the wife blind the husband shout to the wife take me in your back and I will show the way out and they save safe there life . The world as to record nice the new discovery of physic non as something apart from life, they influence and connected link the organic unity that a human being is in mystical union oneness with the organism of the universe , today mysticism is complete update if imply physic otherwise is lame a little expire out of date too , And the world as to drop the Aristotle approach to life which is logical linear the rational association of thought and the double fold logic yes or no or black and white , that was the world mistake, to put down the Socratic approach irrational philosophic and to apply the Aristotle approach limited and shrink , linear logic of rational association of thought and double fold , I suggest to apply the Socratic irrational association of thought which as a seven fold logic yes , no , maybe , maybe yes, maybe not , yes and no together and no and yes together and as all seven color of the rainbow You can apply call the approach of the Hegelian dialectic ,also which as an irrational association of thought of opposite complementary, irrational association of thought and an infinite fold of logic all the color of the rainbow , Aristotelite is a disease of the little men serious and shrink, and he cannot see more then up to his nose because yes and no only shrink denied any further exploration questioning , humanity as remain retarded due to Aristotle, shrink and blind because of the rational association of thought linear logic double fold expire and out of date, because it do not imply further hypothesis for exploration and expanding in questioning , since humanity chose Aristotle thing have gone from bad to worse a great mistake , that as affected all humanity , and sorry all physician of today are affected by the Aristo elite disease and there exploration is slow and shrink, so all the society system are shrink limited because of the Aristo elite disease they cannot see more the three nose , chose a seven fold logic the all color of the rainbow and you are in for a great surprise you will update your life and consciousness awareness in complete different vision and understand approach to live, within reality and without into the objectivity you will expand infinite
........Angelo Aulisa

The new book Mysticism & Physics Chapter 55 Angelo Aulisa

Toward civilization part 15

Hi friends , we continue the journey towards civilization , weapons cannot bring peace violence and bloodshed and killing call violence and bloodshed and killing in a chain reaction infinite , this crazy politicians grave yard diggers think that with weapons wars peace or democracy is brought about , it never work function violence call more violence in a chain reaction bloodshed call more bloodshed in a chain reaction , reaction is not the answer but responding to a situation is the answer , reaction is from your ego mind unconscious , responding is from within your inner being with intelligence consciousness compassion awareness calm centered into your inner being , the attitude of the grave yard diggers politicians up to today to react by building more deadly weapons making more wars bloodshed killing in order to bring peace and democracy and freedom, is propaganda a device to sale more death device call weapons to trade more with death, first of all is a crime against humanity to build device of death call weapons that kill living being sacred holy divine extension of your own consciousness, second wars bloodshed violence killing destruction as never categorically bring any peace , politicians grave yard diggers because they are broker of death device call weapons have reach a point that they are nuclear weapons to destroy 25000 time the world and killing 25000 a man, wen one time is enough and if you added conventional and atomic hydrogen weapons this world can be destroy 50000 time ashamed on you, but have you bring any peace or democracy with all your wars and weapons no only more bloodshed destruction and killing , the politicians have dig a big graveyard of the world for all of you and they still go on using the propaganda device of building more weapons for the sake of money and sale the weapons in a chain reaction , I repeat the way the path is not reacting through the mind ego unconscious no , but responding through your own being with intelligence consciousness compassion awareness to put an end to the trade of deadly device call weapons , and peace is brought about with a bunch of roses and love and compassion and conversation with intelligence and consciousness on peace table on peace deal simultaneous all the nations together because there is no trust but all together simultaneous we CAN , stupid to arm the teachers lucky they are intelligent and they reject the folly delirium of an insane man that he intend to bring peace democracy freedom with violence weapons wars and bloodshed and massacre crazy it has never work it will never work function it only bring about proliferation of violence killing and destruction that by the way for the broker of death as politicians are is ok for them for making more money neurosis of all the world this fucking money stupid , please change attitude and disarm the world right now weapons are a crime against humanity in a new world constitution change production into item beneficial for humanity or face criminal court for crime against humanity . Any how the post that I was writing today regard the Aristotelian approach to life the Aristotelian approach to life is the rational association of thought linear logic of the horizontal dimension of ego mind unconscious from abc to xvz , double fold yes or not black or withe only ,is an infinite regression because you remain always on the surface of objectivity never dive deep within , tomorrow is going be the happiness the love the loving approach but tomorrow never come the happiness love remain always fading away at the horizon receding back at the horizon an illusion and finally death come because death is the only unchanging reality on the surface of objectivity the rest is all changing mutating all the time on the surface of life changing mutating is the only true reality and death the only unchanging reality, what come is today always here ,now flowing in the moment by ,moment the present come gate less gate to the vertical dimension of deeper and higher, of diving deep into your inner being , and that is the Socratic approach irrational association of thought philosophic , of a seven fold logic yes and no, maybe , maybe yes , maybe not , no and yes together, yes and no together , this seven fold logic included the all color of the rainbow , and bring you deep within your inner being Socrates always say know die self know your inner being he bring you within your reality of inner being and he always say and be die self and be your inner being the Socratic approach bring you into the dimension of deeper and higher and be die self means be your inner consciousness he bring you into the higher dimension of universal consciousness and universal body expanding be die self , the seven fold logic allowed you to explore inquire expand into your inner being and consciousness in mystical union with the universal body, Socrates is was the first enlightened of the west a great master,,,,,,,,,, philosopher an irrational association of thought that allowed you to grow to expand and have deep roots into your inner being and consciousness that make a human being healthy whole some , unfortunate the west kill Socrates poison Socrates , and killing Socrates the world the west as rejected the opportunity to live an authentic true life original from your inner being and consciousness , and exploring the inner reality of human being of the organic unity that a human being his , a great mistake to choose the logic linear approach of Aristotelian approach that as make the west and the world blind deaf lame , shrink into the ego mind unconscious into the horizontal dimension at infinite regression . If you ask me I would say to choose the Hegelian approach of the law of dialectic of opposite complementary, which per excellence is the dimension of deeper and higher vertical in fact the thesis deeper and the antithesis higher and the friction and the synthesis of consciousness the Hegelian approach is the approach of NATURE , woman and man synthesis love life can come out of it a living being can be create , life and death synthesis eternal life eternity , heath and love synthesis compassion in an east sense , day and night synthesis time less , hanger and satiety synthesis health peace , and so on so forth micro macro synthesis big bang , and so on so forth I am an absolute Hegelian Nature for me is freedom love spontaneity , the Hegelian approach is an irrational association of thought with infinite fold of logic infinite , actually our language is understood because of the opposite complementary wen you say day you understand day because of is opposite of night when you say love you understood love because of is opposite of heath when you say death you understand death because of life and so on so forth our language is of duality that you like or not is Hegelian of complementary opposite that why you understand the meaning of language , and the Hegelian dialectic law can be apply in all dimension of life of the universe each thesis as is antithesis and friction and synthesis that at its own turn again into thesis and again antithesis and friction and synthesis again , is the approach of nature which is absolutely irrational and the association of thought of the Hegelian dimension approach to life is irrational too , and in the dimension of the law of dialectic you can dive deep into your inner being and rise high at infinite into your universal consciousness simultaneous it seems that Hegel knows already the intrinsic law of the universe as dark matter gravitational energy field which pull together deep within is a gravitation expand everywhere and nowhere in particular into the universal body 24 per cent of the universe, and he knows already the dark energy or expansive energy field that pull up wards higher and higher at the event horizon of the universe and is 70 per cent of the universal body, the real path way is the law of dialectics the law of nature of opposite complementary that include the all color of the rainbow and make of life a mysterious inner and outer journey eternal begin less endless , that the dialectics is the master key if you go with it at infinite in all thesis and antithesis and synthesis you will reach like the ultimate equation of quantum mathematics into infinity infinity that is the law of eternity no begin no end and core and source of the mystery of the universe and of life and death and of all duality , the law of dialectics is a fundamental law intrinsic to the universe the very nature of the universe and is irrational the master key hope you get it or got it in German because this learning will trigger on you a total complete transformation of your life in higher understand quality of living and know die self and be die self .....Angelo Aulisa

Published on February 26, 2018 23:11 • 18 views

The new book Mysticism & Physics Chapter 56 Angelo Aulisa

Toward civilization part 16

Hi friends , we keep the journey towards civilization on and on even if the world generally is going the other way around , the world is in is mission of self-destruction in my understand , they are too many dangerous deadly weapons as nuclear atomic hydrogen weapons chemical weapons etc... that to me are a crime against humanity they are too many wars and division that to me are a crime against humanity , The world is like a broken mirror fragmented in hundreds of piece of different ideology culture philosophy doctrine cult , out of date expire religions superstition and prejudice and pseudo false lies and lies endless that what religions have being preaching down the age , no surprise that the world is in mess chaos in the verge of self-destruction , is the result of lies and neurosis preach down the age by all false pseudo religions , the world is fragmented in hundreds of sect within sect they are big sect like the big shot of religions , and dozens of fragmentation of sects , all in antagonism conflict with each other at the neck of each other at war with each other that create fearless wars destruction bloodshed deaths , and all claim that have the truth bullshit categorically lies categorically all partial interpretation of the little neurotic unconscious serious men , the truth is the empty mirror whole and some the truth is the empty consciousness whole and some no labels no contents no adjectives, consciousness is to infinite to enclose in definition interpretation of the little insane men ancient and expire out of date , consciousness is simple a fundamental law intrinsic to the universal body the very fabric of life call it Boson x because the Boson x is the very fabric of life as consciousness is that keep hold the universe together into an organism , otherwise would have being a mechanical assembled of part and human being would have being robots or machine , but consciousness the Boson confer organism to the universe and the human being is an organic unity link bridge to the universal body through is inner being and consciousness , and consciousness or the Boson confer Mass to the particles atoms matter of the universe , and consciousness is just a pulsation of love intelligence a quality a creativity indefinable just an intrinsic law of the universe and neutral to Gender color race age , is an inner law of the universal body as they are many other inner law of the universe so is consciousness too, the universal body is an unfold of intrinsic law an alchemy of intrinsic law of the universe and so is the law of eternity an unfold of alchemy of inner law of the universe the ultimate canvas of the universe huge bigger vast than universe itself actually were the universe is display and play, and eternity is the ultimate source core of the mystery of the universe and of life and death and of duality, were forms duality time space all annihilate ultimate , but neither the less ultimate alchemy of different play alchemy of intrinsic law of the universal body , The God preach down the age from religions out of date expire pseudo false is an inference an interference not needed and actually do not exist anywhere, is a disease of the little unconscious neurotic men call anthropomorphism that give image character behavior to God they personalize God they give a face to a faceless reality , is an inference interference that create an anthropomorphic men a disease worst then cancer that is on since century enough is enough , is an inference interference not needed dangerous for the evolution growth of human kind is over and over . But neither the less everyone is free to believe in the deception that he chose the believe is a mask to hide the ignorance of the little insane men because no one like to feel that is ignorant, and it do not know a thing , tell me what do you know about god ....nothing at all is an hypothesis at the most an hypothesis pseudo false , even the priest they do not know a thing about God they quote only theory of old out of date expire books but them self they do not know anything , they tell you to pray but what is pray only asking God begging God for favors talking to the sky or to yourself because there is nobody out there , but you are free to hide your ignorance behind an hypothetical fiction as God as believe is for the ignorant , when you get to know through the experience of meditation , silence , love , believe ceases drop end , and so you are free to be disease with anthropomorphism a dangerous cancer that is telling apart the world with fearless wars bloodshed massacred , in short you are free to destroy planet earth and human kind with your childish retarded fairy tale , sorry if we the enlightened one we try to prevent the destruction of the world ha ha , but is just because we live on planet earth and we love planet earth the wonder of wonder of the universe the only green planet that we know where there is blue water air not for so long a guess , but we are try to prevent the disaster destruction done by the barbaric primitive unconscious childish little men insane, because is neurons have being perverted feed with garbage . For Christians the world begin for thousands years before Jesus and he is born out a virgin , the greatest lies ever utter and the resurrection happen body and whole into an utopia illusion paradise, up in the sky nobody knows the location, the resurrection is the most precious item of the mystery of life and death if in this world exist religions is because of death took away death and the all religions disappear, and the Christians give a fairy tale of explanation the biggest lie ever utter because if you go to the cemetery grave all the body are there in putrefaction eat by worm and rats, and the miracles of Jesus a laughing stock for retarded children, and the pray of the Christians and all religions an asking begging requesting of favors please do this give to me that to my wife this and that as if this hypothetical God exist it do not better then you what you need no you know better and you ask beg God, a ridiculous laughing stock approach to life all lies and lies from the begin to the end an absolute hypocrisy , but I repeat you are free to make a full of yourself but not me never and ever, God is a fiction that has never exist it do not exist it will never exist , I ask kindly the resignation of all religions right now , for the good and healthy of human kind , Nice he say God is dead he mistake because God as never being, the idea of God that he as was dead ,but one thing right he say God is dead and men is free that men is free is the greatest truth ever utter , because until God will be there up in the sky an old men with long white bird seating on golden throne on a clouds that is the dramatic idea that human being have of God , men will be only a puppet in the hand of an hypothetical God illusion dictatorial pseudo, we need absolutely to dispose of the fiction of God to be free in first place and to go on into an evolution of intelligence , of consciousness and awareness , ... anyhow I was writing this post to finish my early post on the approach of the dialectic law vertical and multi fold logic, finally it end up into infinity infinity as the ultimate equation of quantum mathematics, the dialectics law is a methodology as is the quantum mathematics , and finally all the irrational association of thoughts and all the folds of logic annihilate into infinity, into non being incorporeal body call it eternity were no time no space no duality no forms survive what survive is awareness I am ness and infinite freedom, bliss peace silence sacred holy divine, and an infinite open relativity and that is the ultimate fold of logic, an infinite open relativity we live the gate door open, or better there is no door no gate at all but just an infinite infinity open, and not absolute at all because the open relativity as no begin or end is just an open sky infinite with no doors or gate, so finally we live the ultimate fold of logic which is not a logic open an open relativity infinite , Jesus say nock and the door should be open to you I say there is no door and not needed to nock as being .always open door gate less, you just let go into the open relativity and you are free into freedom infinite and of course free of the fiction of God that do not exist and will never exist is just an hypothesis fiction for retarded children .....just look around the world the chaos the calamity the wars the killing the destruction the crime, the weapons the disasters all around if you insist that exist a God you would have to give to God all the responsibility, but is not the case because God do not exist and as never exist fortunate and men is free ......Angelo Aulisa

Published on February 27, 2018 10:59 • 25 views

The new book Mysticism & Physics Chapter 57 Angelo Aulisa

Toward civilization part 17

Hi friends. let go on continue our journey towards civilization , because you have trap me now into a conversation that need to be completed even if you are taking too much of my energy and time but let complete the conversation , beloved friends I call you beloved friends and not ship as Jesus was doing the only bigot son of God no sister no brother he was the only bigot son of God megalomaniac, because everybody as the potential to be conscious aware awake enlightened , everybody as a soul an inner being call as better you like in someone is a seed never sprout the majority in someone the seed as sprout and is unfold is quality out wards and the fragrance intelligence is deliver seen , but the potential is in every one without exception , The story of Jesus that he was the only bigot son of God and he do Miracles and he was born out of a virgin are lies mythology to interpretation as mythology , The virgin birth of Jesus was because his Father Joseph was out of station to Erode king , to make a table for him and it took two months and Mary was at home she had a short time flirt with someone , wen Joseph come back he found Mary pregnant and at that time the adultery was punish with lapidating they put the woman into a hole in the ground and they stone the woman to death , so Mary say agree with Joseph that as being the holy ghost to avoid lapidating a tragic death Joseph agree because love Mary , so the holy ghost is a beautiful short time flirt of Mary , the miracles are because there is a competition of all religion to elevate their own prophet Masaya master to an higher stand point so that their own master is the greatest of them all and this make their own master a super hero and them ship basically are tale of mythology anecdote that they point at something else the miracles and the super being too mythology analogy to give interpretation , but the funny things is that all this prophet m Masaya, masters they call there follower ship monks camel it look like a zoo , Then paradise and hell are device of religions to dominate you all they make you scare of hell as little children and what a hell you will be burn forever into fire hell fire what a brutality of punishment so they scare you with fairy tale to dominate you and they make you greedy for an hypothetical pseudo paradise the reward if do they think they say same to dominate all of you with fear and greed but basically it do not exist either paradise that is a wish fulfilment a projection of what you don't get here on earth you get in paradise and what paradise nobody knows the location they Firdausi were you will get boys which are always 16 here the kill the guys in paradise they have boys of 16 always at hand , and river of wine not bottle river here are not allowed to drink in paradise river of wine , and paradise always cool with air condition this the paradise of hot country and so on so forth what you don't get here on earth you will get in paradise, a projection wish fulfill of what you dream here and you don't have , they are no paradise or hell , only device to dominate you all , paradise is a state of inner being of mindless ,were you are not identify with thoughts emotion unconscious staff, no mind, of ecstasy bliss peace silence a state of consciousness universal majestic holy sacred , and hell is a state of ego mind unconscious which you are always identify with thoughts emotion unconscious staff, in anguish anxiety frustration suffering miserable shrink, because of total identification with thoughts emotion unconscious staff and sense this complete identification create hell , here on earth in your life , That is why I say all the time that all religions are expire and out of date pseudo superstition prejudice , because they are real a bunch of lies neurosis without exception that they may have being good for the primitive barbaric unconscious men ignorant because lack of evolution of DNA and consciousness that are link connected intrinsic DNA and consciousness , but totally out of date expire today days for our contemporary of 2018 were we get a men mature emancipate of age that is fed up of lies and lies and want live contemporary with authentic real and in truth , with an update consciousness and awareness enough of lies and make a full of yourself believe in lies the new men want know not believe and known come through experience of meditation silence love , the experience transform believe into trust into authenticity into truth , you don't believe in the sun is there in front of you trust because you see the sun right there so is your inner being through the experience of meditation silence love is right there within you radiating peace love intelligence bliss compassion silence playfulness because an individual center into is inner being is never serious anymore but playful joyful relax calm serene the realization of is inner being link connected him with the all existence simultaneous that means from the known to the unknown to the unknowable and link him with consciousness universal with non being body incorporeal were forms duality time space totally annihilate with awareness that is just a great relaxation I am ness with eternity itself core and source of the mystery of the universe that give to the new men strong roots into is life into is inner being and eternity itself and we get an healthy whole some men , no lies no falsity just a factual reality scientific proved true real the modern men is ready for living through is inner reality and that what he wants , we need a new sunrise dawn of civilization intelligence meditation consciousness awareness right now and the new men will face the sun rise of consciousness alone in silence in bliss ecstasy sacred holy divine without false fake mediators of priest and fake religions are not needed anymore they should find other work to do finish the resignation please of all religions fake and out of date expire right now ....Angelo Aulisa

The new book Mysticism & Physics Chapter 58 Angelo Aulisa

Towards civilization part 18

Hi friends, beloved friends, and so if Jesus is the only bigot son of God, of who we are son maybe son of ...... no Jesus was a megalomaniac , we all born out love and love is the intrinsic essence of your inner being you we are love , in fact love is not an outer experience love you are deep into your inner being, you can share your love wen ever happen share the more you share the more you have it love, the more you hold to share the less you have love , for me is like this what I share my love my experience my wisdom is my forever what I hold and shrink is last forever , for you not for me , anyhow a human being deep within is inner being in essence is love once you realize this basic inner quality of your inner being you are it love and love is an inner phenomenon the experience to share love out words is beautiful wen ever happen share don't be miserable don't hold on but share but remember love is your inner nature and love you are in essence , and Jesus too is born out love as anybody else , the out of date religions expire they make of is prophet always something higher special but is a mistake because so doing they make believe that only the master can be conscious aware enlightened and it is not so everybody without exception as the potential the seed of the soul that can sprout anytime and then the inner being is realize and the seed sprout unfold, the quality flower and love consciousness awareness enlightenment unfold within and without of course the path is meditation done for real, silence love dance painting singing sculpting, any path that take you into no mind ecstasy bliss silence peace love compassion, into universal consciousness into non being were time space duality forms annihilate, and into awareness and the law of eternity unfold begin less endless never born or die, into immortality into resurrection awake from unconscious sleep and enlightenment is yours forever for eternity to come . I am the most democratic man on earth the potentiality of having a soul is of everybody, and so is the potentiality to be conscious aware enlightened , of course enlightenment is not a natural evolution if you wait the natural evolution it will not happen enlightenment is a revolution were your determination afford is require meditation is require love is require silence is require then the revolution but there is no guaranty anyhow it depend on you how much you put on stark your totality is require less than your totality is not ..Enough enlightenment do not come with the parcel of natural evolution but is an inner revolution remember but the potentiality everybody as without exception. Also remember that into consciousness awareness space time forms duality annihilate, and into consciousness you are ageless always fresh sprout the body as a physical age it born and die it live around 80 or 90 years but into consciousness awareness you in essence are ageless or better you are eternal begin less endless never born or die , immortal , you can see with your witness consciousness when you were a child then a boy then young then middle age then old who is the one that see all of your life as a digital color move is you as a witness consciousness, and your consciousness in itself as no age of sort into consciousness you are always fresh sprout. fresh born each moment by moment ageless consciousness awareness are eternal and eternity as no time no space no forms or duality into consciousness awareness you in essence are always fresh born each split second is consciousness that make the generation gap gone transcended , living into your inner being and consciousness and the generation gap annihilate gone because everybody into consciousness awareness is ageless immortal resurrected , no gender no color no age no race consciousness is neutral transcendental above beyond to any interpretation of part of the little men no labels no adjectives no contents, forms age color gender , and anyone can be in mystical union one with consciousness awareness, the link bridge to it meditation silence love that are the path to your inner being, the link bridge to universal consciousness awareness. The span of life that your body as of 80 or 90 years into a space less time less eternity is equivalent to the blink of your eyes and you are gone, so don't waste time to take the opportunity that life is to know your inner being become conscious aware awake from unconscious sleep enlightened, it pass in a split second life is,,,,, very short. So a life lived into consciousness is a life ageless and it will be a better world were the generation gap is bridge transcended were anybody that you interact with is a living Buddha an awakened one an enlightened no matter the physical age , what matter is your inner being and consciousness awareness your been one with it make you eternal ageless ,were everybody and all living being are sacred holy divine in mystical union with the organism of the universe and the human being organic unity of it , A life lived into your body ego mind unconscious a split second and you are death gone and the opportunity that life was last forever, .and you will need a reincarnation a rebirth that existence will go on giving to you at infinite as many body reincarnation you need until you get until you awake to your inner being and consciousness awareness, existence is infinite rich abundant ,until you got you will have opportunity but why not wake up now and live a life of enlightenment of bliss of freedom of ecstasy into no mind, into silence peace is up to you really up to you . But anyhow remember that the situation of the world is going from bad to worse the world is determinate to self-destruction everybody is rushing racing to weapons to get more nuclear missile the invincible nuclear missile ha ha ha, is a crime against humanity dear megalomaniac they are nuclear weapons to destroy 25000 time the world and nations are determinate to destroy each other who knows for what I ask always myself why this confrontation so violent what do you think to conquer, a third world war means the end vaporization of planet earth what you will do after wen they will be no winner, basically, and if you win what you win a planet radioactive contaminated unlivable , the continue building of nuclear atomic hydrogen weapons conventional chemical weapons is a crime against humanity, to kill a living being extension of your own consciousness is a crime against humanity, but it seems that this so call humanity is careless they don't care about civilization that means at priority a world disarm with no weapons of any kind, no this first step of disarmament no civilization will ever happen , civilization means common sense that all and everybody is a friend a beloved friend extension of one own consciousness, on the surface objectivity we are 7 billions of forms but within our inner being and consciousness a single unique oceanic universal consciousness one for the all universal body, and any producer of weapons death device meant to kill holy sacred human being is commit a crime against humanity and human kind, do you want get this or not there is this new awareness consciousness into the world today, small little grave yard digger politicians, broker of weapons death device , the security is bring about with table of peace and disarmament, deal of peace of all nations of the world simultaneous and together, because there is no trust but simultaneous and together can be done we CAN a world without weapons of any kind , and civilization flower a world of friends beloved friends the world existence is not an enemy to you but a beloved friend, open your eyes before that is to late call for conversation of peace and disarmament all over the world and make table of deactivation of all nuclear atomic hydrogen conventional and chemical weapons right now, that is security of Russia of America and of all the world generally , the menace of nuclear weapons is insecurity is the death menace of all planet , abolish the ancient tradition of wars they were doing in the ancient past because they were unconscious barbaric primitive neurotic insane because of lack of evolution of DNA and consciousness they have no better things to do , but today the world as had thousands of convergence of evolution of DNA and great quantum leap into consciousness that the tradition of wars and conquering is over ,today we know what to do celebration playfulness. love dance meditation that create a mystical union with consciousness and universal body and awareness and non being and eternity itself core and source of the mystery of the universe and of life and death and of all duality, and singing painting living in peace love compassion, drop this megalomania of wars weapons military that by the way are childish infantile , I am not scare to me death do not exist to me eternal life eternal journey endless with the body or body less make no difference to me , instead of growing old and die unconscious dear politicians small a little politicians, grow up into meditation , immortality into resurrection into eternity itself, because in essence that is what is a human being, an organic unity, eternity itself immortal resurrected, the resurrection is a conscious alchemy from unconscious to consciousness to awareness into the law of eternity, and not gross physical material , this quantum leap convergence of evolution make you growing up , otherwise you will remain growing old unconscious, and die, infantile childish ignorant . This is my last post I will not write anymore for a while because the book has come to the last chapter, I will publish this book in the next days , has been a pleasure to share my wisdom experience love compassion with all of you beloved friends, and I welcome all of you to read my new book that is on his way to be publish the title I will deliver later thank you all of you I bliss all of you, and hope to have brought about a new understand a new quality to live the life welcome all of you......Angelo Aulisa

Published on March 02, 2018 02:16 • 17 views

www.ingramcontent.com/pod-product-compliance
Lightning Source LLC
Chambersburg PA
CBHW051150220526
45473CB00003B/719